CORNISH CLOCKS

AND

CLOCKMAKERS

CORNISH CLOCKS
AND
CLOCKMAKERS

H. MILES BROWN

DAVID & CHARLES : NEWTON ABBOT

ISBN 0 7153 4999 6

First published in 1961
Second edition 1970

© H. MILES BROWN 1961, 1970

Printed in Great Britain
by Redwood Press Limited Trowbridge Wilts
for David & Charles (Publishers) Limited
Newton Abbot Devon

CONTENTS

List of Illustrations	7
Preface to Second Edition	9
Preface	11
Introduction	13
Clockmaking in Cornwall	15
Cornish Clockmakers	23
Cornish Clocks	35
Cornish Clock Cases	53
References	57
Glossary	58
List of Makers	60
Short Bibliography	94

LIST OF ILLUSTRATIONS

between pages 24 and 25

Plate
I   The ancient turret clock at Cotehele, Calstock
    (*C. Staal, also the National Trust*)

II The church clock at Saltash parish church
    (*A. W. Stilgoe, also the Vicar of Saltash*)

III Wheel-cutting tools used by the Hams of Liskeard
    (*W. H. H. Huddy*)

IV Wheel-cutting tools traditionally used by Caleb Boney
    of Padstow   (*C. Staal*)

V   Clock with month movement by Anthony Ward
    of Truro   (*R. J. Cann*)

between pages 40 and 41

VI   The movement of a typical thirty-hour clock by John
     Belling I of Bodmin   (*D. Haines*)

VII Astronomical clock by Caleb Boney of Padstow
     (*Dr Knowles Boney*)

VIII The restored dial of the Belling clock
     (*D. Haines*)

IX   Plain regulator-type clock by Thomas Odger of Helston
     (*E. Lillicrop*)

X   Eight-day bracket clock by Robert Sholl of Truro
     (*D. Haines*)

A CORNISH WATCHMAKER

Here lies, in horizontal position, the outside case of George Routleigh, Watchmaker, whose abilities in that line were an honour to his profession. Integrity was the Mainspring, and Prudence the Regulator of all the actions of his life. Humane, generous, and liberal, his hand never stopped till he had relieved distress; so nicely regulated were all his movements that he never went wrong except when Set a-going by people who did not know his Key. Even then, he was easily set right again. He had the art of disposing his Time so well that his Hours glided away in one continued round of Pleasure and Delight, till an unlucky Moment put a Period to his Existence. He departed this life November 14th, 1802, aged 57, wound up in hopes of being taken in hand by his Maker, and of being thoroughly Cleaned, Repaired, and Set a-going in the World to come.

> Epitaph in the Churchyard at Lydford, Devon,
> of George Routleigh, 1745—1802.
> Watchmaker of Launceston.

PREFACE TO THE SECOND EDITION

SINCE the publication of the first edition of this book in 1961 interest in Cornish clocks and their makers has continued. A good deal of additional information has come to hand augmenting our knowledge of the makers in the county, and in presenting this enlarged edition the list of makers and others in the trade has been extended to include those up to 1870.

In particular the recent research into the early Cornish years of the famous craftsman John Arnold, improver of the chronometer, and inventor of some features still to be found in modern instruments, is reflected in the list of Bodmin makers. It is gratifying that the Bodmin Old Cornwall Society, with the co-operation of the Borough Council, has taken up my suggestion and erected a plaque recording the residence in the town of this eminent man. Fascinating is the hypothesis which emerges that the Arnold and Belling families, so close in the juxtaposition of their places of work, were associated in business.

Many people have written since 1961 describing clocks in their possession, and specimens have been brought along to talks I have given on the subject. For all this a debt of gratitude is due, and also once again thanks must be expressed to Mr H. L. Douch, of the County Museum, and to Mr P. L. Hull of the County Record Office for their amplifications of the original list, drawing on Cornish wills and other sources. And I am grateful to Miss Elizabeth Johnstone for permission to quote from the Trewithen MSS, to Mr Frank Michell for the loan of a scarce Directory, to Mr W. G. Negus of Camborne for permission to quote a letter of John Hocking, and to Mr Clive Osborne of Clerkenwell for encouragement to produce a second edition of *Cornish Clocks and Clockmakers*.

St Winnow
1970
H. Miles Brown

PREFACE

SOME 40 years ago a small boy was gazing at the first grandfather clock he had seen. Placed at the top of the stairs it seemed a grand and impressive sight. An interest was kindled in that moment which waxed and waned and waxed again through the years, fanned by friendship with a watchmaker at Saltash, whose workshop became a kind of Mecca. Among the bits and pieces and tools many processes of the art and mystery of clockmaking were expounded.

Later years brought ministry in Cornwall, and the consequent friendly opening of cottage doors enabled the collecting of Cornish makers' names to begin. Thus the present little volume was conceived, through which it is hoped that the interest will be handed on, and that the craft of these old men, forefathers of Walter Visick, my childhood watchmaker hero, will receive the honour due to it.

In writing such a book as this the author risks a fall between two publics. There are those who want an interesting and chatty tale of the men and their work, with a list of makers in which they can discover the date of the household clock. And there are the experts whose desire is to have everything set out clearly in horological terms.

It is the hope of the author that the two will find something of value—that perhaps the casual reader may be led on to study more seriously the history of horology in all its fascinating by-paths, and that the expert will gain an enhanced respect for the craftsmen of these far shores.

Such a book as this could not be written without gathering

a great number of debts of gratitude, for help and advice given. Especially am I grateful to Mr H. L. Douch, Curator of the County Museum, for his interest and helpfulness, particularly for many extracts from early newspapers; without his assistance this book could not well have been so complete. My gratitude must also be expressed to Messrs K. Parsons and J. K. Bellchambers for advice and criticism and the exchange of ideas; to Mr W. H. H. Huddy, of Liskeard, for interesting information about the old makers and their methods, and the opportunity to inspect old movements in his wonderful workshop; to Messrs W. H. Coulls and C. Hewlett for the loan and gift of books; and to Mr J. H. Pethybridge for allowing me access to the notebooks of John Belling, which have added such an interest to the tale of the makers. Thanks are also due to Mr P. L. Hull and his staff at the County Record Office for their ready production of documents and facilities for making notes.

But above all I am indebted to those many clock owners who have allowed me—usually with a word of apology for the dust on the top of the case!—to inspect their clocks and take notes of them. Many owners answered press articles in which information was asked for and without their help this book could not have been attempted.

Such a work as this cannot in the nature of the case be free from error, neither can it be complete. The help of readers in correcting the errors and amplifying the list of makers and their clocks will be appreciated by the author. And local historians may well feel impelled by the many gaps in this survey to unearth yet more information to illuminate the lives and work of those worthy men who adorned their own district long ago.

H. Miles Brown

Truro

INTRODUCTION

THERE are few old-time crafts whose products can vie in interest with the clocks of a bygone age. They make an appeal to many who do not know much about them, and would wish to know more of their intricacies, and of the men who made them without machinery, save of the simplest sort.

These clocks carry with them something of the personality of their makers, their skill and resourcefulness, their whims and even their humour. And in addition they have perhaps been 'in the family' for a long time, closely identified with the significant moments of successive generations, binding the one to the other by a kind of symbol of permanence. As infants each generation watched the slow beat of the pendulum in its cobwebby case, or followed with wondering gaze the rise of the heavy weights at the Sunday winding. In courtship, marriage, and the last moments of life the familiar dial has been scanned for the significant hour, and the old clock has given to each generation the same service, impersonal, but yet something more.

The science of time measurement has of course left these clocks, and indeed mechanical devices of all kinds, far behind. For those who are interested in the present state of horological practice there are many standard books wherein the mysteries of modern time-measurement are revealed. Similarly the many attempts to mark the passing hours which our ancestors evolved before they hit on clockwork mechanism of the familiar sort are well provided with descriptive literature.

In between the ancient and the modern lies the English

clock, which has for a long time received its due share of the serious attention of specialists. But with the greater popularity of bygones and antiques since the late war destroyed so much of beauty and value, a more general interest has been taken in the horological past. A wider public is aware of the masters of the English craft, or at least of the London scene. Names such as Fromanteel, East, Tompion, Graham, Quare, Knibb and Clement are no longer the passwords to a closed company of experts; they have come alive as the acquaintances of the many. And the way in which the English clockmakers and chronometer makers came to lead the world in the accurate measurement of time, and in the construction of beautiful pieces of furniture, is everyday knowledge.

In the past the focus of attention was not unnaturally on the makers working within the small limits of the City of London and its immediate environs, over which the full influence of the Clockmakers' Company, founded in 1631, was felt. The best craftsmen were to be found there in profusion. In consequence the name of a London maker on the dial of a clock gives it a merit often greater than the piece deserves. In comparison the makers in the provincial towns and in the countryside were, until recently, passed by and undervalued.

There are plenty of signs that this lack of balance is being redressed. Serious and rewarding interest is being taken by horological students in the practice of the craft in centres such as Bristol, Liverpool, Wigan, Derby, Leicester, Halifax, Newcastle and other places. In the same spirit a survey of regions, or of well-defined areas has been under way for some time, and competent researches have been completed on the clockmakers of Wales, Yorkshire, Leicestershire, and so on.

It is at this point that the justification for the present work can be seen, and into the general picture can be fitted the story of the Cornish clocks and their makers.

CLOCKMAKING IN CORNWALL

The Equation of Time

The Fifteenth of April
    and Seventeenth of June remember
August the Thirty-first
    and Twenty-fourth of December
On these Four Days
    (and none else in the Year)
The Sun and Clock
    both the Same Time declare.

                      From a printed label of c.1760.

AMONG the earliest examples of clocks made in England are those in the West Country Cathedrals of Salisbury and Wells, probably constructed at the end of the 14th century.

In Cornwall the first clear traces of the making of clocks appear in documentary sources, and it is significant that the earliest record of a clock in Cornwall would seem to be that of St Mary's Church, Launceston, the most easterly and Devonian of our towns. The Borough Accounts from 1432 onwards mention various expenses about 'le clokke', quite clearly a time-keeping mechanism. Only half a century therefore separates the well-known clocks of Salisbury and Wells from the first parish clock of Cornwall.

Such primitive constructions—one very early example still exists in East Cornwall, as we shall see—would be the work of smiths and locksmiths. Only gradually did the craft of clockmaking become distinct and in the provinces there was much overlapping in the case of larger and tower clocks. The making of domestic clocks for use in the home did not it seems begin in England until about the closing years of the 15th century. For long only the richest could afford the luxury of marking the time in their homes, and there would be few such in Cornwall.

In various records of about the middle of the 17th century the names of a few Cornishmen active in clockmaking appear, and though it is probable that some of them worked on larger timepieces rather than on clocks for the home, there are one or two who are termed 'watchmaker' in the description of their craft, which would in the usage of the period indicate the smaller domestic clock rather than the pocket watch of the present day.

It was in this direction of the domestic and finer clocks that full scope was given to skill and to art. The formative years of the London tradition of clockmaking, and the finest period of English craftsmanship lie between say 1670 and 1750. Up to about 1670 the makers were feeling their way in this more delicate field. New ideas multiplied, such as the anchor escapement and the 'royal' pendulum, rack striking work, wheel-cutting engines and the like. And after about 1760 the lines on which the English makers were producing clocks became more or less stereotyped.

In the provinces, and especially in such a distant area as Cornwall, there would tend to be a time-lag, and the best period of provincial clockmaking comes a little later, say up to 1800, at first alongside and finally succumbing to, the trend towards standard productions. Late 18th-century movements throughout the country are largely similar as forgings and castings from central warehouses became more readily available. The mark of the maker remains, of course, but the interest lies in the occasional unusual movement and in the surviving traces of an individual approach to the craft.

From the middle years of the 18th century makers increased greatly in number up to the opening decades of the 19th century, though as we shall see the craftsmen actually capable of making clocks tended more and more after about 1790 to use whole movements made out of the county.

In estimating the place of the Cornish clockmakers in the general English scene it has to be remembered that the county only provided a poor market for a great part of the period of active clockmaking. Makers were in trade primarily to support themselves and their families, and to carry on the business of their fathers. They could not afford expensive decoration, or the time needed to produce complicated movements except for the small minority of wealthy clients. It should be remembered that a considerable proportion of the population of Cornwall—which was about 100,000 in 1688—was engaged in mining and ancillary trades. The monthly reward of their labour would seldom be greater than £2. A tinner working on 'tribute' in 1778 could clear 30s monthly, with the chance of more; a 'captain' might receive 40s. A market for clocks was not found among such, and indeed the working miner was often without means of telling the end of his period of labour in the mine right up to the 1860s, when watches were comparatively cheap. Nor was the wage of the agricultural worker or of the fisher folk greatly different.

The clocks of the earlier period were costly in comparison with the wages of the many. It is to the gentry, the mine-adventurers, the beneficed clergy, the wealthier farmers, that we must turn to find the market for clocks up to the first half of the 18th century. The price of clocks can be illustrated thus. In the Buller records an inventory made in the month of June 1689 of the goods at Morval House, near Looe[1], includes in the best hall 'one clock and case £2'. An inventory made in 1755 of the estate of Nathanael Hender, of Lanteglos-by-

Camelford, includes—'in the passage a clock and case, £1-10'. From the revealing little notebook of John Belling, who was active as a clockmaker in Bodmin from about 1706 until 1761, it appears that he usually charged £3 3s for a 30-hour clock and case, and £4 10s for an eight-day clock of superior sort. Even a large-size house would not have the number of clocks usual today. One or perhaps two would suffice for a large farm or mansion, the striking being sufficiently loud to be heard throughout the building. The inventory for the Caerhayes Castle sale in 1789 mentions only two clocks. One stood on the stairs and the other in the kitchen; traditional places in many houses to this day. For poorer would-be owners there was perhaps a clock-hire service. The will of John Pye, of St Columb, clockmaker, proved in 1740, includes a note of a clock 'out upon hire', valued at £2.

It may be of interest to glance in passing at the prices charged by makers in other parts of the country at the same period. In such a similar provincial area as Wales, for instance, prices were very much the same. Samuel Roberts of Llanfair Caereinion, Montgomeryshire, priced his clocks from £2 to £6 6s, though the majority sold for under £3 10s.[2] At a later date—the opening of the 19th century—the Hebdens of Wensleydale, Yorkshire, were getting no more than £5 for a longcase clock which took them a fortnight to construct. But the best London makers could command as much as £19 for a quite ordinary eight-day clock in the mid-18th century.

From Cornwall, therefore, as from other remoter parts, ambitious and forward-looking craftsmen tended to move to places where their work would be more rewarding. For instance, John Arnold, one of the perfectors of the chronometer, after being apprenticed to his father in Bodmin, left the town and ultimately settled in London, where he became established. It was the son of a Cornish immigrant who made the first wholly-American chronometer in 1812; he was William C. Bond, a good county name.

As the 18th century wore on a market developed in Cornwall among the new middle class. From the 1770s with more efficient methods and means of power an impetus had been given to mining in Cornwall. Greater numbers were employed, fresh skills emerged. The complexity of mining processes called forth specialist groups, such as engineers, captains, count-house clerks and the like. The beginnings of industrial organisation among the mining adventurers demanded clerks and accountants. At the same time the more effective use of land, and a combining of fishing skills, helped toward a more prosperous era with repercussions on trade of all kind.

A wider market for the clockmakers' wares was provided by these groups, and the retail traders who supplied their daily needs. In addition there were the factories, count-houses, workshops, foundries and offices where they worked, and where a clock would be a convenience if not a necessity. Cook's Kitchen Mine, Illogan, paid Mary Bennetts £5 8s in 1790 for an eight-day clock, presumably for the count-house. The old Levant Mine count-house clock stands at the present time in the mineral room of the County Museum, Truro. What scenes of conviviality it must have witnessed at the audit dinners may be left to the imagination.

It is in the last half of the 18th century and the first years of the 19th that the greatest concentration of Cornish clockmakers occurred. The Census of 1831 recorded 83 watch and clockmakers at work in the county in that year. They were of course found principally in the towns, such as Truro, Bodmin, St Austell, Penzance, Liskeard, Falmouth, Launceston, St Ives, and Redruth. Their close association with the mining and engineering industries can be traced in the lives and work of William Wilton, of St Day, who made survey instruments for the mine engineers, William West, model-maker to Richard Trevithick at Hayle, and so on. And the maritime connections may be inferred by the frequency with which Cornish clocks possess a moon and tide dial. It may be guessed that some of these were scanned at times with anxious eyes to see if things were propitious for a run of smuggled goods!

Profound in its effects upon the industry here as elsewhere was the emergence of cheap methods of production outside the immediate district. Ways of diminishing the considerable labour of making clocks had long been sought, and as far back as 1670 Robert Hook, that inventive genius and Curator of Experiments to the Royal Society, had evolved a wheel-cutting engine which vastly reduced the tedium of cutting clock wheels and ensured their accuracy. Increasingly during the decades which followed, specialisation developed rapidly. Firms sprang up in the larger cities where partly-prepared material could be had. The similarity over wide areas of such things as hands, spandrels, and the like shows the increasing use being made of these supplies. The work of the actual makers lay more and more in detail, finish and ornamentation in the second half of the 1700s, though of course many men did continue the use of locally-produced items until it became uneconomical to do so.

From about 1790 it was possible to order complete movements from firms in London, Birmingham and other centres. At first no doubt the use made of this convenience would be small, but it seems that makers increasingly turned to these sources for their lower-grade work. A great proportion of the 'grandfather' clocks of the ordinary sort will be found to possess behind the painted dial (occasionally this is also met with behind a brass dial) an intermediate cast-iron rectangular frame pinned to the front plate of the movement. This is usually taken as a sign of a late and factory-produced movement. The rectangular frame made it possible to attach the feet of any dial without interfering in any way with the assembled movement. The frames often carry the names of the ironfounders producing them, such as Wilkes, Baker; Finnemore & Sons; Wilson, Birmingham; Osborn, Birmingham.

The production of factory-made movements was not mass-production in the modern sense of interchangeability of parts; it was rather batch manufacture by cheap labour under one roof. But it was inimical to true craftsmanship. It soon became practicable to sell clocks, locally fitted with a cheap case in ordinary materials, at a price within the reach of the greater number, but unprofitable to master makers. From about 1810 even villages began to boast of 'clockmakers', selling their wares to the lucky miner, the tenant farmer, the thrifty cottager and others who doubtless imagined that the name of the dealer on the dial denoted an actual maker. Places such as Altarnun, Morwenstow, Kilkhampton and many others produced ready dealers. Some of these were willing to combine the selling of clocks with the vending of general village necessities. From contemporary advertisements we find clocks in company with groceries, ironmongery, drapery, and in one case (Jonathan Hodge, Helston) coffin furniture!

When clockmakers of the old sort were forced into a variety of lines, it was as a defence against a new threat. This was the influx of very cheap American clocks from the early 1840s onwards. These were produced by machinery in bulk in factories, and at a price within the reach of everyone. The *West Briton* in the 1840s long contained a weekly advertisement by Ramsey of Devonport offering these clocks. They were also advertised in 1846 by Charles Levy of Truro, and Morris Hart Harris of Penzance, who offered them at 17s 6d each. In the columns of the same paper can be seen the effects of this influx, in the frequency with which businesses were sold up on the death of the proprietor rather than being carried on by the son.

Nevertheless, clocks were made in the old way, albeit by a dwindling number of men, up to the end of the period under review. Family tradition, and the personal recollections of the last generation, tell how the old men were seen at work. The survival of their tools also assures us of this. W. Hotton Huddy of St Newlyn East and later at Tregony was working as a maker as late as the 1850s, for example.

A similar state of things was found in other parts of the provincial field. In 1916 the owner of a clock by Robert Chasty of Hatherleigh, Devon, reported that he had been allowed when a child to see the old maker at work cutting the teeth of the wheels. This would be in the 1840s. In Yorkshire the manufacture of clocks by hand also went on till the 1850s at least, while in Wales, according to the careful research of Dr Iorwerth Peate, there were makers active until the 1860s.[3]

We may judge that the art of clockmaking had pretty well ceased in Cornwall by about 1860. An occasional clock may have been made by some apprentice bent upon proving his skill, or by some craftsman for his own enjoyment, but it was the rare accomplishment. The name 'clockmaker' today usually indicates merely a skilled repairer or a resourceful retailer.

Enough will have been said on this score to make it plain that it is impossible, in the last part of our period, to be dogmatic as to how far any specimen clock was actually made in the county. The line between dealer and maker is not easy to draw. In many instances it is probable that men capable of constructing clocks would turn to the factory items when pressure of business or demand for cheap wares made it necessary; on other occasions the wheel-cutting engine and the lathe would be set to work once more.

CORNISH CLOCKMAKERS

> Sacred to the Memory of Caleb Boney who died November 12, 1826, aged 79 years deservedly celebrated in this County for his proficiency in Astronomy, Mechanics and Music.
>
> Headstone in Padstow Churchyard.

Up to the present over 150 names of Cornish craftsmen at work before 1800 have been recorded, and a somewhat larger number for the following half-century, making a total of just over 500. Of course, finality in this search is not likely to be attained, since there will constantly be new names and hitherto unreported clocks coming to notice from obscurity. But as complete a list as may be of the makers in the county up to 1870 will be found at the end of the book. It is hoped that it will be useful to clock owners and collectors, and also to others interested in a county craft of days gone by.

It will suffice in this chapter to select from the number a few typical makers about whom something more than ordinary is known, or whose work might illustrate the general trends which have been already reviewed.

The most informative source of knowledge concerning the skill and capabilities of any maker must obviously be surviving clocks of his make. From the details of the dial and hands, the layout of the mechanism, the occasional touches of ornamentation on hammer-stops, rack-tails and the like it is usually possible to glean some indication of the date.

With one or two exceptions however there are no surviving clocks by Cornish makers before about 1700. We are thus thrown back upon documentary sources for evidence of the early workers on mechanical timekeepers. And the first references to clockmakers in the county will be found in the archives of parishes possessing clocks at a remote period.

It has already been mentioned that St Mary's, Launceston, had a clock in the year 1432. It was kept by John Nevhalle, the sacristan, who was paid annually the sum of 5s for this double duty. In 1450 we find the sexton was John Dawe, also responsible for keeping 'le clokke', and in the 1460 accounts there is mention of Walter Robyn and payment to him for setting 'le clokke' with the time; in a later entry he is referred to as a 'maker'. He was in all probability a smith, who had learned the skills of clock construction and maintenance.[4]

Similarly at St Austell the clock, whose stone 24-hour dial yet remains on the west front of the tower and may be attributed to about 1480, is traditionally said to have been the work of John Austle, a miller. Other churches early possessing clocks were Bodmin (about 1500), St Columb (about 1585) and Liskeard (1604). The Cornish writer Carew mentions in his *Survey* 'old Veale', of Bodmin, who in the late 16th century was a self-taught wheelwright, carpenter and clockmaker, as well as surgeon and chemist. But it may be surmised that the clocks he made were few.

In 1657 there died at Ladock, near Truro, William Randell, who is described as 'watchmaker' in a rhyming epitaph which will be found on a tablet on the south respond of the tower arch in the church there. 'As hee made watches, soe did keep good watch.' The terms 'watch' and 'clock' were at this time overlapping in meaning, and it is possible that some simple form of clock may have been constructed by this man, perhaps an amateur, since he is described as 'gentleman'.

James Fonnereau, of Truro Lane, St Gluvias, is another whose brief mention as a 'watchmaker' arouses our curiosity.

I The ancient turret clock at Cotehele, Calstock (page 36)

II The church clock at Saltash parish church—a typical 'birdcage' movement, *c* 1730 (page 37)

Wheel-cutting tools, now in the County Museum (page 40)

III Used by the Hams of Liskeard, early nineteenth century

IV Traditionally used by Caleb Boney of Padstow

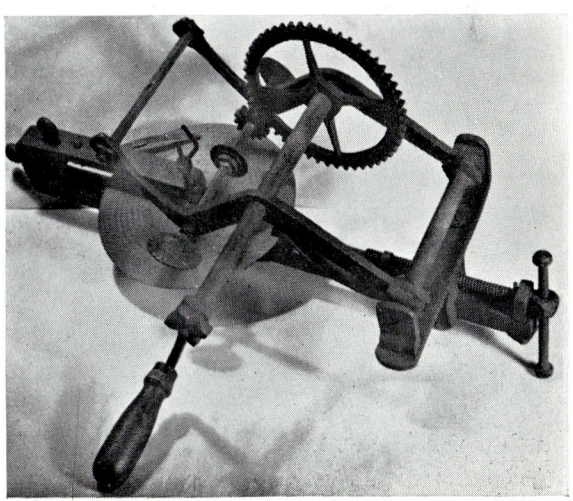

V Clock with month movement by Anthony Ward of Truro, *c* 1705
(page 41)

CORNISH CLOCKMAKERS 25

This time it is a passing reference in a deed of 1676.[5] There were well-known watchmakers of this name at La Rochelle, at the same period. This man may have been a relation, and perhaps a fore-runner of those many Continental craftsmen who have found refuge in the West Country from political upheavals abroad.

The Borough Accounts of Truro record the instruction to James Anthony of that town, a local maker, to alter the mechanism of the public clock from presumably a foliot and verge escapement, 'He makinge it a pendilow', in 1699.

From a comparatively early date in our study Bodmin, the county town, seems prominent in the Cornish horological scene. It is tempting to find some link with the old skills fostered by the many trade guilds known to have existed there in former times. The self-taught Veale has already found passing mention. Another Bodmin maker was Christopher Hawke, or as he signs himself, 'Hocke'. He was paid the sum of £3 10s in 1671 by the Mayor of Lostwithiel for repairing their clock and making a dial for it on the tower of the church. The agreement exists between Hawke and the Mayor of Grampound, Peter Hearle[6], for the making of a new clock for the chapel of St Naunter in that borough. The agreement is dated 20 November 1673. The Mayor is to have the old clock and pay Hawke £6, as well as supplying cords and 'peases' (i.e. weights) for the new one, which was to strike the hour on the bell. It is just possible that this replacement of the clock marks the adoption of the anchor escapement and pendulum, though there is no evidence of this.

Among the many famous makers who advanced horological standards during the 17th and 18th centuries were at least two who sprang from Bodmin stock. John Ellicott, senior, was apprenticed to John Walters in 1687, his parents having removed from Bodmin to the metropolis perhaps to take advantage of the greater skills and scope to be found there. John Ellicott served as Warden of the Clockmakers' Company from 1731 till his death in 1733. More eminent still was his son, also John, born in 1706 and elected F.R.S. in 1738. He invented a form of compensated pendulum, and popularised George Graham's cylinder escapement for watches.

The famed horologist John Arnold was born at Bodmin in 1736. His father was a watchmaker in the town, and the boy was apprenticed to him, till after a quarrel he left home and made his way into Holland. Later young Arnold settled in London and attracted attention by constructing a tiny repeating watch which he had set in a ring and presented to George III in 1764. From this moment his future was assured.

Arnold turned his attention to the problem of exact timekeeping at sea, which was occupying the best horological brains of the time. He succeeded in constructing chronometers of greater accuracy and more economically than those of Mudge and Earnshaw, which were themselves superior to the first successful instruments of John Harrison. Arnold was awarded a share—after much bickering—of the Government award for the best chronometer, though his son, John Roger, taken into partnership in 1787, actually received it.

These eminent makers, Ellicott and Arnold, do not seem to have returned to Cornwall at any subsequent period, and they do not therefore truly figure as Cornish makers. But no story of Cornish clockmaking would be complete without mention of them. It is interesting that a passing reference to 'Mr Arnold'—John's grandfather, Richard—occurs in the notebook of John Belling, and that an Ellicot signs as witness to the will of John Belling II in 1806.

The Bellings of Bodmin were a line of makers who remained in the town, working between about 1706 and 1850; their present-day descendants manufacture in Middlesex a well-known brand of electric cooker. The founder of the clockmaking line was John Belling I, who died about 1761. He was an excellent maker, and supplied a wide area with clocks, mostly 30-hour and eight-day longcase, but including one or two small turret movements, as that at Treworgey, St Cleer, dated 1733, and a charming bracket clock with quarter repeating work, about 1760. He becomes a human figure for us, perhaps the most fully limned of all the Cornish clockmakers, through the survival of a little jotting-book which he carried during the years 1737-54. It seems likely from the number of clocks attributed to John Belling that he employed several people and possibly some outworkers. From his notebook we can see the man at work, and on his occasional journeys to Bristol by sea—possibly sailing from Padstow—to purchase from the warehouses there material for the clock trade. He buys files, chain wire, clock and watch hands, emery for polishing, and what he spells as 'crewsibols' for casting brass. He brings home some sharkskin for Mr Arnold; a bottle for his father-in-law; a coffee-pot for himself. We see him setting off to clean, repair or regulate some clock in a nearby farm or manor house: '4 April 1738. Clean'd Mrs Oliver's clock at St Clears and Rec'd 4 shillings in full.' He takes orders for clocks, noting whether they are to be 30-hour or eight-day, the height of the place they are to stand in, the size of the dial and its pattern.

He also stands revealed in these casually-made notes as a receiver in a small way of smuggled goods. On Thursday, 10 September 1751 he notes he has 'Two bottles of French brandy in ye Cubbord Conseal'd', and others were hidden away at various dates up to the following March.

He charged £3 3s for a 30-hour clock, and £4 10s or £5 for an eight-day clock. This, as we have seen, was a considerable sum for all but the few in the Cornwall of those days.

Belling's trade went as far afield as St Just-in-Penwith, where he sold Mr John Maddern two eight-day clocks, Tintagel, Port Isaac, Mevagissey, Lostwithiel and St Cleer. In some cases he took part payment in kind. In 1739 Mr William Parnall of St Austell desired a clock, evidently from its price a plain 30-hour. On 9 November Mr Parnall gives him in cash £1 1s, and also 21 lb 3 oz of black tin valued at 8d the pound, towards the clock. Later Mr Parnall brings him some old 'Grav plats' and more tin, and presumably has his clock. In connection with this transaction it is interesting to note that John Belling often cast the spandrels of his dials in tin.

He was succeeded by his son John II, who died in 1806, a man of substance, leaving to his son John (III) his tools and some property, and to his daughters further property. The sons carried on the clockmaking trade, but the general appearance of clocks made by them is more commonplace. Painted dials had come into fashion during their lifetime, and the interest and value is less.

Other makers in Bodmin were the Broads, Richard and his sons John and Joseph, who traded alongside the first Bellings. Their clocks are without features of interest, except that it is on record that they sometimes delivered their clocks by hand, on one occasion tramping 12 miles over the moor to a lonely farm, carrying the clock between them over the miry bottoms and past the wondering sheep.

As time went on, Bodmin dropped out of the foreground of the picture, giving place to Truro, which in the 18th century was becoming the fashionable centre of the county. It was the time of development through the mineral wealth, the building of the town houses of the great mining promoters, the era of Lemon, Daniell, of Wolcot and Polwhele. There had been good makers at a very early date. A fine month clock of a standard at least equal to that of many a London maker exists, having been made by Anthony Ward of Truro about 1705; it will be described in the next chapter. Ward was paid 2s 6d in or about 1705 for putting a clock right at Luney, in the parish of Creed, and also £5 7s 6d for a new watch.

The two Anthonys, James and his son William, were in business up to about 1750. The former, as we have seen, was responsible for the 'modernising' of the Town Clock in 1699, and the latter is represented by a few competent eight-day clocks which show signs of careful local manufacture.

It is with Richard Wills, however, that we reach the zenith of the Truro skill, and perhaps indeed that of the county. He began work about 1750, and was making clocks up to about 1805. Wills in his day speedily became a fashionable craftsman, if the quality and distribution of his work is any guide. One of his clocks is illustrated in a standard work on English domestic clocks.[7] It has a tall slender case with a bell-top, and a silvered dial with an arch in which are displayed the then fashionable automata, in this case a man sawing and another planing, the action being taken off the striking train. Another, preserved locally, shows 'High Water at Truro Key', and yet another by Richard Wills exhibits the mechanism, unusual in provincial clocks, known as 'maintaining power', which keeps the clock running forward when it is being wound, and thereby avoids the otherwise inevitable loss of half-a-minute or so during the winding process. Wills also made the first clock for the new tower of St Mary's Church, Truro, in 1770. He made sundials, and constructed, or partly constructed, an astronomical clock with a musical train which was completed and exhibited by his son William. It is described in the next chapter. The younger Wills, born in 1756, was a man of ability and ingenuity like his father. His business included the making of all kinds of clocks, and the repair of watches, clocks and organs. For the barrel-organs found in country churches he was able to supply new tunes. William Wills died in 1819, and the firm came to an end having been in trade 70 years 'in the principal thoroughfare'.

Other Truro makers of repute included Robert Sholl, born 1765, who made the beautiful bracket clock in the Truro Museum, and Philip Polkinghorne, died 1801, one of whose clocks in a well-proportioned mahogany case will be familiar to patrons of the old Red Lion hotel in Truro, where it stood in the foyer.

While Truro not unnaturally would seem to have risen to pre-eminence in the quality of the craftsmen's products, other towns had their interesting makers. Those of Falmouth, especially Carlyon, Nancolas, Cumming, Eva, and Martyn, were without doubt of a superior sort, and their products carry a smack of the sea, as we might expect. Of these men, Richard Eva should be singled out. He was born about 1734, and in January 1782 married Margaret McDowell, being described in the Falmouth parish register as 'watchmaker'. A longcase clock with a planetarium in the arch is attributed to him, and there are extant in Cornwall some good longcase clocks bearing his name, and a bracket-clock which will be described in the next chapter. It appears that Eva worked for several years up to 1780 at Tregony, and clocks still in the district bear this town's name on the dial instead of Falmouth. Another has in the arch a tide dial calculated for the port of Falmouth.

Concern with 'those that go down to the sea in ships' also appears in the role Eva played in partnering Edward Cook, a painter, of Falmouth, in patenting 'an Apparatus, Machinery, or Instruments on a New Construction, for the Purpose of Taking Observations and Altitudes, both by Sea and Land, without any Dependance on the Visible or Sensible Horizon'. The patent is numbered 2087, and dated 9 February 1796.

CORNISH CLOCKMAKERS

The basis of it is a kind of vertical segmental mercury-level. Richard Eva died in 1806, his obituary notice describing him as having been 'for many years a respectable clock and watchmaker in that town'.

In Redruth was Moses Jacob, whose period of activity terminated with the 18th century. In addition to being a competent maker, he was one of the first to deal in mineralogical specimens from the Cornish mines. His interest to us here is enhanced by the fact that he employed as a journeyman one James Dawson. This man had been a French subject, had served in the army of the King of France and after the Battle of Fontenoy had deserted to Prince Charles of Lorraine, who let him go where he would. He chose to go to Vienna, and wrought in his trade of clock and watchmaking, which he had learned in Dieppe from his father.

After working as a journeyman in several places in Germany, the Low Countries, London, Wales, Ireland, and the West Country he came to Moses Jacob. In 1769 he contracted to serve this man for two years at a wage of 16s every week, excepting Jewish festivals and holy days, Jacob being a practising Jew.[8] In January of the same year, Dawson had married Grace Pascoe, a widow of Redruth. After the unusually varied experiences enjoyed by this man, one might have expected him to have set up on his own, but of this there is no trace.

Moving further north in the county, we meet one of the most versatile and widely-known of Cornish clockmakers, Caleb Boney of Padstow. A published note at the time of his death in 1826 states that he was born at St Teath in 1747, and had little or no formal schooling. After teaching himself to read and write Boney found employment at Delabole. From that village he went off to Liskeard to work in a solicitor's office, but was frequently in trouble for reading in bed at night. Returning to North Cornwall he became a self-taught carpenter at Camelford, and then turned to clock and watch making at Padstow, which remained his residence until his death in November 1826. Two other Calebs, a son and a grandson, were in the same line of business. The former went bankrupt in 1828.

Several elaborate timepieces and astronomical clocks were made at Padstow by Boney, some of which are still extant and are described in the next chapter. There are many more ordinary clocks made by him in existence. Boney was also something of a musician, much in demand for tuning instruments, and from about 1806 he included bell-founding among his many pursuits. Dunkin, in *Church Bells of Cornwall*, 1878, traces four bells bearing the name Boney in Cornish towers —St Minver, Sancreed, and two at Sennen. He also appears in the Kenwyn, Truro, parish accounts as journeying from Padstow to see the bells there and to give an estimate for their new casting. At the age of 78 he rode the 41 miles to Hayle to tune bells recast at the Hayle Copperhouse Foundry for Helston Church.

After Boney's death the schoolmaster of St Merryn, one Chapman, blossomed into 'verse' in his praise—

> Sun, moon and planets he did make
> Like to Sir Isaac's notion
> And neatly fixed them to his time
> Which set them all in motion.

In the far west of the county clockmaking skill was allied to engineering ability in the person of William West, 1751-1831. He started his career as proprietor of a smithy at Helston, and evidently also engaged in clockmaking in the same period, since a good longcase clock by him, of about 1775 and bearing the town name 'Helstone' on the dial, is reported from America, whither it was recently carried from

CORNISH CLOCKMAKERS 33

Cornwall. West came into contact with John Harvey, like him a smith, and the man who in 1779 established the foundry at Hayle which was renowned throughout the 19th century for its products.

William West married Joanna, one of Harvey's daughters, at St Erth in 1784, and thus had as his brother-in-law Richard Trevithick, who was then experimenting with high-pressure steam engines. West is described as having been a sensible steady-going mechanical man, though somewhat obstinate. As a craftsman in brass and small forgings he was the man for making models of Trevithick's inventions, and the first model high-pressure engine was made in bright brass by West in 1796 or 1797. He shared with Trevithick and Andrew Vivian the patent granted in 1802 for a steam carriage, and did much work on the engine, in Cornwall and London.

The financial success of these ventures was disappointing, and about 1808 West set up as a clockmaker and instrument maker at St Ives. His association with the Foundry remained close until his retirement in 1828. Several clocks by West are still in existence, one displaying some ingenious variations on the familiar 'Adam and Eve' movement in the arch of the dial. Popularly known as 'West's chronometers' these clocks had a high reputation for accurate timekeeping.[9]

In a few cases the commencement of clockmaking in the county has led to the establishment of a line, father succeeded by son or younger brother, and so on. The line of Belling in Bodmin has already been mentioned. The firm was in existence right up to the end of the period covered by this survey—1870. Another old-established line springs from Joseph Ham of Liskeard, who was making clocks for the district from about 1780, and whose descendants and associates have been in business in the same town up to the present time. It appears that the early Hams turned out mostly 30-hour clocks, made up in batches. A few at a time would be put on a farm wain and hawked round the farms and cottages of the countryside. Other makers and dealers were supplied with movements and clocks, on the dials of which would appear the name of the retailer, and not that of the Ham firm. The wheel-cutting engine, chainmaking tools and other gear used by the Hams survive and are still usable.

A name frequently occurring on the clocks of the northern part of the county is 'Reynolds'. The towns which appear on the dials of Reynolds clocks are Padstow, Wadebridge, Launceston, Egloshayle, Newquay and Lewannick, this last being a mere village. The earliest of the Reynolds would seem to be Thomas, who in 1786 took a lease of a dwelling-house built by Humphrey Gink on the east side of Wadebridge, 47 ft by 14 ft; he is described in the Wadebridge accounts as 'watchmaker'.

The middle years of the 19th century saw an influx into this country of foreign clock and watch makers, preponderantly of German origin, as a result of political upheavals on the Continent. Several of these men found a home in Cornwall; their names will be easily recognised in the list of makers. Forerunners of this influx were two brothers named Beringer, who in 1830 or thereabouts emigrated from the Black Forest region of Germany and settled in Cornwall. At first they worked as travelling clock repairers, later setting up individual businesses at Penzance and Helston. This was about 1832. Later still, other branches were opened and partnerships entered into, but there was no actual clockmaking done in the county by this firm. The interest lies in the example of early trading in foreign movements, as some from the Black Forest manufactories, with the typical wooden frames and bodies to the wheels, appear behind painted dials bearing a Cornish name as the towns of origin. The firm is still in existence.

CORNISH CLOCKS

January ye 4th, 1739—
  Clocks to make as under

for Ths Hill a 30 howar clock with howar and minut
for Henry Chelew a minut day of month Clock
for Mr Pearn a 30 howar day of the month & minut clock
for Mr Hoskin of Dewstow a playne 30 howar clock
for Mr Jno Raw of Camelford a minut and day of the month clock.

From the Notebook of John Belling I.

THE passing references in the various records do not give us a great deal of information concerning the clocks to which they refer. The Launceston clock, it may be inferred, was a striking clock since the first entry relating to it in the Borough Accounts for 1432-3 includes a note of payment for wire for the clock bell. This clock was new made in 1466. In 1469 is the entry 'springwarde for le clokke'. It would obviously possess a weight-driven movement, and it or a successor was responsible for a fatality in 1654, the parish Registers recording for 10 April of that year the burial of Julian Gliddon, 'who was slain with a pease of the great clock'.

Hidden away in a maze of lanes in the parish of Calstock on Tamarside is the medieval mansion of Cotehele, now in the possession of the National Trust but for long the residence of the Edgcumbe family. In the chapel, which was licensed for Divine service in 1411, there remains a very ancient turret clock of unusual interest. It has only recently become known to the horological world.

This clock is representative of a small group whose movements are not contained in the more usual 'birdcage' or 'four-poster' frame, but arranged vertically. In this case the striking train of two wheels and a fly is planted above the going train. This latter consists only of the great wheel, revolving once in every hour, and the crown wheel. The verge still possesses its original foliot, which is underslung, swinging at the bottom of the movement. This is a unique feature so far as known clocks of this sort are concerned. Similar vertically-planted movements survive here and there, but the total number of those known is very small and most have been converted to pendulum control. The clock at Sydling St Nicholas, Dorset, bears the date 1593.

The Cotehele clock never had a dial, and struck the hour on the old bell still in the turret. The wheels are all of iron, and the rims are separately made, being held within the split ends of the four spokes, which were then closed over; yet they run truly. Two of the striking train pinions are of the lantern style, comprising pins held between two plates. Both these characteristics of the wheel construction indicate an early date of manufacture, but expert opinion is chary of committing itself too firmly to an estimate of the age of the clock. It is now accepted, however, that it may have been made as early as 1490. The clock may be seen by the visitor to Cotehele, and is in going order, having been recently overhauled and restored.

The pendulum was introduced into ordinary English clocks about 1660. The anchor escapement was invented some 11 years later. At first many clocks, both of the tower kind and the smaller domestic ones, had a pendulum fitted to the verge escapement which was so familiar to the makers. The Truro Borough clock was so altered in 1699.

An interesting reference to another clock in the transitional stage occurs as a result of John Smeaton's interest in the aftermath of a thunderstorm which happened while he was engaged in the building of the Eddystone Lighthouse in 1757. This storm badly damaged the church at Lostwithiel, and in an account of the effect of the lightning which he sent to the Royal Society we find a reference to the wrecking of the clock.

> The verge, that carries the pallets, was bent downwards as if a ten-pound weight had fallen upon it. The crutch that lays hold of the pendulum looked as if it had been cut off by a blunt hook. . . . As to the pendulum which hung pretty near the wall, the upper part was struck with . . . violence against the wall.[10]

Smeaton was one accustomed to the proper use of technical terms, and we can therefore be certain that in 1757 the tower clock at Lostwithiel had a verge escapement and a pendulum. It may indeed have been the very clock which Christopher Hawke repaired in 1671 and furnished with a 'Watch or Dyall upon the tower', though the pendulum would no doubt have been added later.

A movement with the anchor escapement, more typical of the early 18th century, remains at work in the tower of St Nicholas & St Faith Church at Saltash. It was constructed in the reign of George III, about 1730, by a local or journeyman clockmaker. It has an iron four-posted frame with two trains which run for about 48 hours. The hours only are struck. The trains are mounted vertically, side by side, in the frame in the usual fashion for these movements. The pendulum beats seconds-and-a-half, and is a lengthy affair. No maker's name appears anywhere on the clock, which may originally have had but a single hand indicating the hour.

These turret movements are not uncommon, but many have been replaced with more modern mechanism, while the old 'works' have been allowed to rust away. The old movement of the Tregony town clock, a very similar movement to that just described, did duty as part of a pigsty fence, and an attempt to rescue it was just too late; it had a few days before gone for scrap. It bore the name Richard Eva.

A rather later movement is that traditionally once in St Mary's Church, Truro. It is on record that the first clock for the new tower of this church was made in 1770 by Richard Wills of Truro, and did duty until 1851, when it was replaced. This old movement which is preserved in good order on private premises locally, could well be Wills's work. The trains are planted vertically; the frame is very plain.

Several of the older houses in Cornwall possess turrets in which clocks exist, some dating from the early 18th century. They are often smaller than the church tower clocks, since the room which could be allotted to them was limited and the striking mechanism does not need to be so massive. Some of these clocks, as those at Calenick, Truro and Trewithen, Probus, indicate hours only. These smaller clocks could be constructed by makers who ordinarily worked at domestic clocks. There is an example at St Cleer, dated 1733, made by John Belling of Bodmin.

The domestic clocks throughout the period we are discussing are mostly longcase, or to use the popular term, 'grandfather', clocks. Bracket or mantel clocks by provincial makers are not so common, though there is a handful by Cornish makers which will find mention here. It has been suggested that the provincial makers' difficulty in obtaining the large springs necessary for these clocks is the reason for their avoidance. Frequent replacements would no doubt be called for.

From the middle of the 18th century a refinement of the longcase clock is found in the 'regulator', in which everything is sacrificed for accuracy of time-keeping. The escapement is not the ordinary anchor, but a kind of deadbeat, normally

Graham's. There is no striking train, and a compensated pendulum is fitted to reduce errors caused by change of temperature. The whole movement is solidly and carefully made, while the case is usually plain and functional. These regulators were constructed to serve where accurate timekeeping was desired, such as in observatories and factories, and for the regulation of common clocks in the workshops of the clockmakers. They are therefore less often seen than the usual longcase sort, but there are one or two by Cornish makers at which we shall glance.

One lantern clock, with the usual turned corner pillars and finials, is known to the author. It is enclosed in brass engraved sides with applied chapter ring and alarm disc. It bears the name 'John Belling, Bodmyn', and shows the hours only. The movement, dating from about 1710, is 30-hour.

Longcase clocks do not bear any marked county characteristics in the dials or movements, but are akin to the styles found in the whole of the south-west of England. This is what might be expected, when the recourse to provincial centres for tools and materials is borne in mind. With the supply would go the latest idea and fashion. It is true that the provinces would tend to lag behind the London styles by a decade or more, and that they would be locally interpreted rather than slavishly followed. In the later 18th century the longcase clock went out of fashion in the metropolis, but was enthusiastically continued in the rest of the country. Local styles long remained similar, but differences developed at the end of the century.

In the northern parts, Lancashire and Yorkshire for example, cases became wide and dumpy, dials were overengraved and legibility reduced. But in southern England the slender proportions of the old London style persisted. In Cornwall, as we shall see, casework was not always good, but except in the later and cheaper clocks for cottages, where the height had to be kept down to fit under the low ceilings, the proportions continued to be sightly and in some instances handsome.

Before describing individual domestic clocks which illustrate the various trends, or display some unusual features, it will add vividness to our study if we pause to glance at a typical workshop of the first half of the 18th century. This we are able to do, since according to tradition which seems likely, the Belling premises still remain unaltered, and in addition there are Belling's notes of his tools, which, with the inventory of goods and tools attached to the will of John Pye, clockmaker of St Columb, proved in 1740[11], will give ample scope for reconstruction.

The Belling premises consist of a loft over a stone-built lower storey (where the furnace would have been) about 18 ft by an average width of 15 ft. The front was all of glazed casements. In this small factory, perhaps with some outworkers, a great number of clocks was produced, of which many survive. A variety of files was needed constantly, and these we see in the lists of goods bought by Belling, and among the tools left by Pye.

Wheel-cutting engines were of course widely employed. That belonging to Boney still survives; it is a simple rough affair, but capable of cutting teeth of sufficient accuracy. A later engine, once belonging to the Hams of Liskeard, also survives, and is a superior machine still in working order. Casting-pots and a furnace for the melting of brass for the dials, plates, and wheel-blanks, shears for cutting chain wire, vices, anvils, a simple lathe or two, grindstone and oil stones, materials for polishing and finishing, all feature in the scene.

Shortly before his death John Pye had just begun a 'chaine clock'—probably a 30-hour movement—which was valued at

VI The movement of a typical thirty-hour clock by John Belling I of Bodmin, dated 1756 (page 43)

VII Astronomical Clock by Caleb Boney of Padstow, *c* 1810 (page 47)

VIII The restored dial of the Belling clock (see Plate VI); note Cornish tin spandrels (page 43)

IX Plain regulator-type clock, eight-day, with dead-beat escapement
by Thomas Odger of Helston, *c* 1800 (page 49)

X Eight-day bracket clock by Robert Sholl of Truro, *c* 1790, now in the County Museum (page 51)

£1. Two unfinished jacks were valued at £2; and two clocks and cases not finished, £5. Stocks of iron wire for chains and pendulums, brass scrap for melting down, stocks of clock and watch hands, springs, glasses and keys find mention in Belling's book. Pye's will refers to a 'little roome' containing a small vice, and perhaps we may imagine some industrious apprentice learning there to file out delicate parts to the satisfaction of his master. It is with such simple implements that the craftsmen of the day turned out those complicated mechanisms which still delight us.

A few clocks by Cornish craftsmen may now usefully be selected as illustrative of various trends in design. The first Cornish domestic clock which deserves to be mentioned is a fine longcase example by Anthony Ward of Truro. It has a month movement, with the extra wheel in the train needed for the longer run, and heavier plates to carry the weights. In this clock these latter weigh 23 lb each, and are brass-cased, always a sign of superior workmanship. The dial is typical of the opening of the 18th century (the clock was probably constructed about 1705). It is square, with applied silvered chapter-ring and seconds ring, on a gilded main plate matted finely at the centre. Around the central arbour is engraved a rose motif. The chapter ring bears fleurs-de-lys between each figure, quarter-hour divisions on its inner edge, and $7\frac{1}{2}$-minute divisions are also marked. The two winding holes are ringed, and there is the usual square opening for the circle showing the day of the month. The border of the dial is engraved with a herring-bone pattern. The movement is beautifully finished, with a locking-plate striking train, the count-wheel being on the back plate. The spandrels display the crown and cross supported by cherubs. The case is of Spanish mahogany, a very fine specimen, but it is almost certainly a later replacement. The hands are original and well-cut.

Thomas Wills of St Austell, who died in 1739, is represented by an eight-day clock in a case with a lacquer or japanned finish. There are two or three by this maker with this sort of decoration. The date of the first-mentioned clock would be about 1720. Here the dial carries the arch at the top, which became popular from about 1710 onwards. At first it commonly carried a 'Strike/Silent' dial, or a raised boss with the maker's name and town, or a motto such as 'Tempus Fugit'. As time went on the arch was used to display automata, or a moon and tide dial. On the clock by Thomas Wills of St Austell it carries the maker's name. The movement is rack striking, which allows the hours to be repeated. The chapter ring is not silvered, the plain brass finish to this ring being more commonly found than the silver finish in the provinces except for the finer specimens. This dial by Wills carries the inner quarter-hour divisions as well as the almost universal minute-ring and five-minute numbers. The curl of the 5s and 3s is not brought round to the body of the figure, and this again is a sign of early construction, since in later dials the curls become more pronounced.

The inner divisions are a sign of a date previous to about 1750, when they appear on a clock with a minute hand. They are found less frequently after that date, as they served no real purpose after people were familiar with reading a dial with two hands. A good clock from the date at which these markings have just disappeared is one from the hands of John Troughton of Helston. This man kept the clocks of Dr William Borlase, Rector of Ludgvan, from 1735 to 1743. Again the clock is an eight-day with rack striking movement. It has an arched dial carrying the maker's name and the town in which he worked, in the arch. The case is the original one, of oak, tall and well-made. The hood has small side-panels of glass, as indeed many Cornish clocks do. There was a school

CORNISH CLOCKS

of thought which ascribed these only to London clocks, but there are plenty of examples in the West Country. The date of the clock by Troughton is about 1750.

John Belling I made more clocks of the 30-hour than the eight-day kind, and a typical clock by him has a very sturdy, carefully-made movement with an occasional touch of embellishment, e.g. on the hammer-stop. Striking is regulated by a locking plate which appears on the back upright of the lantern-like frame of the movement. Country makers often used this form of construction rather than plant their wheels within plates, as in the eight-day clock. Later makers of 30-hour clocks, however, did sometimes place the mechanism between plates, as may be seen in a specimen by William Clode of Camelford, made about 1755-60.

Usually the dial of a Belling clock is 10 or 11 in. square with spandrels of an early cherub kind, often cast in Cornish tin instead of brass. The centre of the dial will be finely matted. The name 'John Belling Bodmyn' appears in small cursive engraving at the lower edge of the chapter ring. There may be inner markings to the chapter ring for the quarter-hours, though these were on the way out in this maker's later years.

Some makers, including John Belling I, constructed 30-hour single-handed clocks, though the central minute-hand had long been familiar by the middle of the 18th century. These single-hand movements are often taken to be older than their actual date of manufacture. The purpose of this simplification was twofold: clocks so made were a trifle cheaper, and in addition they were more easily read by simple folk. They continued to be made in Cornwall until about 1750. John Belling was certainly making them, as he records, as late as 1740, and Thomas Wills of St Austell, James Sellick of Marazion, William Vian of Liskeard, and probably others, obliged their customers in this way.

It will be seen how important the dial details are together with the movement and general layout in estimating the date of a clock. The tendency as the century proceeded was for both spandrels and hands to become more open. The wavy form of minute hand increasingly occurs after 1770. The centre of the dial was often engraved with floral motifs of formal pattern, or with simple scenes, such as ships, cliffs, churches, cottages and mansions. While many makers continued to make and engrave their own dials, they could be supplied possibly by travelling engravers, or from central depots, either as blanks or prepared and engraved. This sometimes produced amusing results. There is one such dial, of about 1790, which bears the name 'John Reynolds, EGLESLAIL'. This version of the place-name Egloshayle is not likely to be perpetrated by a Cornishman of that district. A similar mistake appears on a dial bearing the name John Bennett, with High Water at 'Kenequie Cove'. It is at least possible that these and other dials were ordered from distant factors. Such suppliers existed as Charles Blakeway of Albrighton, 1774-1795, who advertised 'wheels cut for clockmakers at 6d per set, and Dyal Plates engraved at 2/6 each'. The Yorkshiremen could get their dials in 1804 from Whittaker & Shreeme of Halifax.

The pattern of spandrel generally sold seems to have satisfied many of the local men, as we do not often find that they designed their own. The same pattern crops up time and time again, and the total number of different styles is surprisingly small. Later, designs became more and more meaningless, often twists and curls without form, affixed to the dials rough without the touch with the chisel and the finishing methods employed by earlier makers.

Many clock movements display the date of the month either through a square opening at the lower part of the central dial space or in a semi-circular opening with a pointer, behind which moves an engraved disc. Both of these forms of mechanism require to be adjusted at the end of the short months, as perpetual calendar work is most unusual and no old Cornish specimen is known to the author. The larger date opening—sometimes it is very large—is more typical of the West Country. It later tended to become smaller but still segmental. In Cornwall these styles parallel those of the rest of the western counties.

Fashions changed again in the last two decades of the 18th century. Sometimes the five-minute figures on the chapter ring were reduced to the quarters only, as with some modern clocks. George Routleigh of Launceston produced a dial of this kind, about 1790. From about 1775 the engraved and ornamented brass dial with applied chapter ring gave way in cheaper clocks to the painted dial, either square or arched as the traditional sort. The spandrels were of course not of metal, but painted designs, usually floral, though sometimes romantic scenes or representations of the four seasons were used instead. These dials were manufactured and the designs applied by stencils at central depots. Arabic numerals for the hours occasionally occur on later dials.

Another style which appeared in the last decade or two of the 18th century is the flat silvered or plain brass dial, with no applied chapter ring or spandrels, but instead engraved figures, with motifs or scenes. The dial circle may consist only of a circle of dots marking the minutes. Here again there may be an arch to the dial. Such features are found on clocks by Richard Behenna of Penryn, Philip Polkinghorne of Truro, Richard Eva of Falmouth, Roger Wearn II of St Erth, Joseph Veal of Mevagissey, John Bennett of Helston, James Cumming of Falmouth, John Pope Vibert of Penzance and many others.

At this later time it was common for the arch, both in

painted and traditional styles, to carry moonwork, with a revolving plate painted with moons which waxed and waned behind semi-circular shapes on the dial, to imitate the moon's varying phases. In a sea-conscious county such as Cornwall, it would be natural that the dials should also carry tide indicators, and as the tides depend on the moon, it was simple enough to attach a semi-circular dial calculated for the time of high tide in the place the clock was to serve. Such places as Truro Key, N.E. Hayle Key, Plymouth Key, Mount's Bay, Trewavis Cove, Falmouth, Padstow, St Mawes, Penzance, and Kenneggy Cove, together with others, appear on the tidal dials of the county, and no doubt were of use when tide-tables were hard to come by.

Another favourite attachment was a simple kind of moving figure in the arch of the dial. A rocking ship is quite common, as it is easily worked off the escape arbor by a bent wire, supporting the ship against the background of the arch, usually painted to represent the sea. Clocks by Roger Wearn II (St Erth), Henry Curgenven (Helston), Moses Jacob (Redruth) and others display this feature. Or there may be a hovering bird; in one specimen by John Bennett of Helston it is two, a hawk and a buzzard, moving successively over a frightened-looking chicken! There are also examples of a face in the arch with moving eyes (Anthony Nancolas of Falmouth), a feeding swan (William Wills of Truro), or a scene from the Garden of Eden, with Adam and Eve continually stretching out their hands to the forbidden fruit. In one such example of this movement, by William West of St Ives, the serpent twists himself around the tree as the clock strikes and Eve reaches out with each blow of the hammer. Richard Wills of Truro and James Cumming of Falmouth each constructed clocks with men at a saw bench and grindstone in the arch, actuated by the striking train.

Clocks of this period (1750 onwards) rarely have any great horological interest. Numbers of makers were producing competent eight-day clocks all very much alike in the little workshops of provincial towns. Some made up movements for others to sell in their own name, as did the Hams of Liskeard and J. P. Vibert of Penzance, as well as selling many themselves. The beginnings of multiple methods can thus be seen even within the county itself, though as has been explained, much of the supply came from more distant sources.

The interest henceforward will lie in the unusual or complicated movement, and there are a few which will merit description here inasmuch as they come from Cornish makers. There are for instance astronomical and musical clocks by Caleb Boney of Padstow still extant. One such musical clock is an eight-day movement with a third train actuating a large barrel studded in the usual way with pins. There is a nest of twelve bells housed at the top of the movement and struck by the hammers moved by the pins in the revolving barrel. In the arch of the plain 13 in. brass dial a steel hand can be turned to the title of one of the four tunes which the clock will play before striking each hour: i, Money Musk; ii, St Bride's Bells; iii, Song; iv, CIV Psalm. The tunes were not in full order at the time the clock was examined. The case is very large, of mahogany, and the clock would appear to date from about 1800. Across the dial is inscribed 'Caleb Boney, Penzance', and presumably that maker had a retail establishment there as the movement is full of Boney's characteristic touches.

Of the astronomical clocks by this maker two known are almost identical in their layout. It may be of interest to quote a description of one of them, supplied by the present owner who is a direct descendant of Boney's. It has an eight-day movement in excellent working order.

The outer chapter ring is divided into 24 hours of the day and night, and each sub-divided into twelve, so that each sub-division represents five minutes of time. The age of the Moon chapter ring, lying in the same plane as the hour ring, is divided into 29.5 equal parts and carries a steel hand to point the time on the hour circle.

On this steel hand, the wire portion supports the gilt motif representing the sun. The other steel wire carries a small ball representing the moon, one half being cream, the other half black. This ball rotates, thus showing the phases of the moon.

Within and a little below, is a plate showing on the outer edge the calendar months and days of the year, allowing for the 28th of February, but not for Leap Year. The Signs of the Zodiac are also shown.

On the first plate, but further in towards the centre, the Ecliptic, Equinoctial, and the Tropics are laid down, as well as all the stars of the first, second and third magnitudes, according to their right ascension and declination, those of the first magnitude being distinguished by eight points, the second by six, and the third by five.

Over the middle of this plate and a little above it is a fixed plate marked Earth, around which the sun moves in 24 hours and 50.5 minutes and the stars in 23 hours 56 minutes and 4.1 seconds.

The ellipse represents the horizon of the place the clock is intended to serve, and all the stars seen within the ellipse are above the horizon at the time.

The other clock is similar, but not at present working. They are both well-made and the dials are very beautifully engraved and silvered. The name Caleb Boney, Padstow, appears on both.

Other unusual clocks by Cornish makers include a fine regulator by Martyn of Falmouth, made about 1800. It stands in a good mahogany case, and has a circular silvered dial with centre seconds hand, dead-beat escapement and compensated pendulum. The dial is inscribed 'Martyn's Regulator'. It is in perfect order and the movement is good. Another regulator

CORNISH CLOCKS

of a rather coarser sort comes from the hand of Thomas Odger of Helston. This possesses a plain dial of the ordinary kind, and has a dead-beat escapement. There is no striking train in either of these timepieces, nor in the two astronomical clocks by Boney. The astronomical and musical clock completed by William Wills of Truro in 1817 deserves mention, though its present whereabouts, if it survives, are unknown to the author. Wills's advertisement in the *West Briton* of 29 August of that year describes it fully:—

> The front exhibits a circle on which are marked the hours from one to twenty-four, and which, with the minutes, are pointed out by indexes from the centre. Within this circle is a bisected plate, shewing a section of the Horizon—the Sun is represented by a gilt ball, which by a compound motion rises, sets and shews the altitude of the real luminary in the Heavens, throughout every day in the year, in this latitude.
>
> On the bisected plate are inscribed the name and longitude of several of the most remarkable places on the earth; by which the hour at these places is constantly pointed out.
>
> In the upper corner of the plate a small spherical body shews the continually changing phases of the Moon. The opposite corner shews another globe revolving on its axis, and exhibiting the diurnal motion of the Earth.
>
> On a plate in one of the lower corners are inscribed several concentric graduated circles, by which are pointed out the Dominical letter, Golden Number, Epact, Cycle of the Sun, Leap Year, etc. In the opposite corner is a circular plate and index, shewing the time of high water at the principal sea-ports in Europe every day.
>
> There is also exhibited a small Orrery, regulated on the true principles of the solar system, having the Sun in the centre, whose place in the Ecliptic is pointed out every day throughout the year. The several planets are represented by ivory balls, which exhibit their true places in the Zodiac, and their revolutions, in the most accurate manner. The ball representing the Earth is inclined to the plane of the Ecliptic, in an angle of $66\frac{1}{2}$ degrees, and thus exhibits the phenomenon of the seasons.

The Moon is also similarly represented and moves round the Earth thirteen times, whilst the Earth revolves round the Sun.

Behind the Solar system is represented a clear sky, in which is a figure of Apollo seated on a cloud. Every four hours this figure appears to strike a lyre, and beat time with its feet, whilst an organ plays several tunes.

This clock, which is a handsome mahogany case, will be disposed of by way of subscription in 120 shares at one guinea each, the money to be paid at the time of subscribing. Should it be previously disposed of by private contract, the money will be immediately returned to the Subscribers. Any Lady or Gentleman, who will favour the owner with a call, will be shewn the same gratis.

By their humble servant
William Wills.

Remarkable as the clock must have been, and sincere though the praises of 'the best judges who have viewed it' undoubtedly were, the requisite 120 guineas were not forthcoming. Wills died in the January of 1819, and the business passed to his sisters, who disposed of it in 1820, the clock list still not full, in spite of repeated advertisement.

There is a charming little bracket clock movement by John Belling I, with a quarter repeating train, actuated by pulling a cord, whereupon the last hour and number of quarters is repeated on a nest of six quarter bells and the hour bell. The movement has a verge escapement, and the spring barrels are connected to the fusees by chain, though this may not be original. The back plate is plain, the case is modern. The dial is a small edition of one of Belling's usual longcase dials, 8 in. with arch, and in the arch is a boss inscribed with the maker's name, 'Jno. Belling, Bodmin'. The date would be late in the working period of that maker, say 1760, just before his death. It is in good working order still.

There is a good bracket clock by Richard Eva of Falmouth, dating from about 1795, with an eight-day spring movement

CORNISH CLOCKS 51

with striking train. The escapement is a verge, which in bracket clocks was common until about 1800. This type of escapement was not so susceptible to slight out-of-level placing as the anchor kind, and so was still made for this type of clock when it was long obsolete for longcase pieces. It was useful in clocks designed to be carried from room to room when most houses had but one or two clocks.

Robert Sholl of Truro, who died in 1815 aged 50, is represented by a very good bracket clock of about 1790, better cased than the one by Eva, but possessing a similar verge movement. Without actual dismantlement it is impossible to be certain, but it would appear that the movement is by the maker whose name is on the dial. The clock by Eva has been so dismantled, and no indication of 'foreign' making is to be found, so we may be pretty sure it is from his hand. The Belling clock plainly exhibits characteristics of that maker, so we may be more definite. All these clocks are superior specimens and a credit to the county skill.

We may close this chapter on Cornish clocks with a brief description of a curiosity, which unfortunately bears no name. From its history it would appear to have been made in the county, and some of its features remind one of the work of Caleb Boney. It could possibly be one of his clocks, but no definite decision can be made on the evidence available. Or possibly it may be from the workshop of Joseph Pedler, smith, of Withielgoose, who was a maker of 'curious clocks', according to a printed family history, and born about 1739. The clock seems to have originated in the Bodmin-Padstow area. The case is a normal longcase one in mahogany with an arched hood surmounted by a pediment broken by curved horns of the style 1780-1800. The centre of the dial is a 12 in. circular plate painted white. Around the edges it is divided into six sections, each in turn numbered in five-minute spaces up to 60. The central part of this circular plate carries a spiral trough of two complete turns. In this trough lies a marble. The plate revolves once in six hours, so that the minutes can be read off from the edge of the plate by a pointer. As the disc revolves the marble is carried up the trough, in the lowest part of which at any moment it rests. At the hours are the Roman numerals painted opposite where the marble lies. At one o'clock the marble is at its highest, near the middle of the disc. There is a hole there, connected by a tube to the other end of the spiral. The marble drops down to the commencement of its climb, to repeat this every 12 hours.

The clock possesses no striking train, and has an eight-day movement. It is wound by a handle whose revolutions are carried to the barrel by an idle wheel. The wheels are light, but the plates sturdy. The pillars are cylindrical and not turned down between the plates. The train is planned for the peculiar action, and the disc shows careful manufacture, the tube and marble being counterbalanced. The hood carries a thin plate surrounding the revolving disc, and in the arch is a dial graduated up to 60. There is no action now connected with this dial, and it is difficult to see how there could have been any. The clock keeps good time, though the marble tends to move in jerks. As it only indicates the hours this is not a grave fault.

It would be easy to dismiss this curious clock as a late toy, but for the fact that the case and movement appear to be old and may be tentatively dated 1800.

The reader will have had before him ample illustration of the various trends in design. A general pattern of development has been traced. Within it however the individual approach of each maker and the occasional touch of ingenuity will be easily recognised.

CORNISH CLOCK CASES

Supplied by J. HODGE, HELSTON,
>Ironmonger, silversmith, and dealer in
coffin furniture.
>>From a note found in a Cornish clock.

So far the cases in which the products of the clockmaking workshops of Cornwall were housed have found little mention. This is partly due to the fact that the making of cases was of course a branch of cabinet-making, and most clockmakers would have employed others to construct their cases for them. The general succession of case styles may be followed in any standard work on English clocks, as the best Cornish specimens do not display any county characteristics.

For the sake of completeness, however, readers may wish to have in the briefest form some reference to the fashions and materials and finish common to particular periods, though in the provinces the greatest variety is found, and styles persisted long after they were extinct in the metropolis.

The earliest longcase clocks date from about 1670, but as we have seen there are none known of this early date in Cornwall. The characteristic material of this opening period is ebony or lignum-vitæ veneered on oak. The moulding under the hood would be convex until about 1700, and the hood would slide up in grooves to give access for winding and regulating the clock. The top of the hood would have a portico form. After 1700 cases would tend to become taller, though in Cornwall they were always limited by the low-browed ceilings common to many old buildings in the county. The emphasis on height rendered the lifting of the hood impossible, and it was arranged to slide forward, in the familiar way, with a hinged glazed door in front of the dial, opening outwards to allow the hands to be reached for adjustment, and for winding. The convex moulding under the hood gives way to the concave, while the door of the trunk remains square at the top. A few cases are found with a lacquer or japanned finish, not often of a high standard in Cornwall but pleasant by way of variety. Thomas Wills of St Austell seems to have used this style frequently; the surviving clocks by this maker appear to be housed still in their original cases. Apart from this, oak, or walnut veneer, are also found in the first half of the 18th century, the former being by far the more usual in Cornwall.

Mahogany cases are rare in the first half of the century. As the wood was not imported in any quantity until 1745-50, it only became a fashionable material for furnishing in the latter half of the 1700s. From this time the corners of the trunk exhibit a chamfer, which later becomes a quarter column, often reeded. The hoods may follow the outline of the broken-arch dial, or be built up to a flat top with an oversailing moulding. After about 1770 a fashion common elsewhere appears in Cornwall, with the hood showing a curved pediment, roughly in the form of a bell with the front and back flat, the front exhibiting shallow carving or fretwork with three ball finials at the corners and top.

From about 1780 hoods also display commonly a finish with broken pediment and swan-neck horns, and there are numerous variations on these simple themes. The pierced side-panels, finished either with a fret or with glass, enabling the movement to be seen, are common in the county; the glass circle in the door of the trunk to display the swinging

CORNISH CLOCK CASES

pendulum is, however, rare. The doors cease to be square-topped and tend to follow the outline of the hood, or to be fashioned into wavy outlines, and later to display 'Gothick' arches and points.

There are some very fine specimens of Cornish cases, housing first-class movements, and of a standard equal to good provincial work elsewhere. The town makers naturally had more exacting standards and wealthier clients. Richard Wills and Philip Polkinghorne of Truro housed clocks in excellently-made and well-proportioned cases of mahogany, both of a similar bell-topped design. Anthony Nancolas of Falmouth used a flat-topped mahogany case with fret under the top moulding. John Troughton of Helston supplied a clock in a well-proportioned oak case with a flat top to the hood over an arched dial; the capitals to the pillars at the sides of the hood are especially good. James Cumming has a clock with an arched hood topped by a flat moulding and fret. The case is mahogany with a touch of carving under the hood moulding. Peter Hitt of Liskeard has a clock with a mahogany case, the hood having swan-neck pediment and the door of the trunk banded with tulip-wood. Many other worthy examples could be adduced, as numbers of makers employed, for their better work at least, casemakers who were expert in this specialised form of cabinet-making.

It must, however, be admitted that numbers of Cornish clock cases are very poor, both in material and design. They are often made only of soft-wood, painted or grained, and not seldom badly finished. In many instances the present cases are replacements, either ancient or modern. It is an interesting speculation as to why even good movements were so poorly housed. It may have been poverty on the part of the client, or the absence of true cabinet-making skill in an age and area where there could have been little work to keep many employed. Or it may have been due to the sheer abundance of rough timber available at the mines, and the ready handiness at carpentry possessed by many.

The names of few case-makers have come down to us. It is on record that the Hams of Millbrook and Liskeard employed a man almost whole-time making cases for the clocks manufactured by that firm; his name was Willcocks. Inside the case of a clock whose movement was supplied by Beringer and Schwerer of Redruth is a pencilled note, '1850, John Shepherd, maker, Redruth'. But these are very late examples. We do know that the first Belling clocks were put into cases which cost about 12s to 18s.

Occasionally one comes across a case which, if not beautiful, has a tale to tell. There is a Bristol movement in the County Museum, Truro, with a case made very passably by the mine-carpenter at Levant Mine, and in the same town is a longcase clock whose case was made, according to family tradition, by Captain Will Richards, about 1810. He was the last of the notorious Prussia Cove smugglers, and we can imagine the surroundings in which the clock was first set going there and the use to which its moon-phase work was put.

The tale of the average Cornish clock, however, will be somewhat more humdrum. Most are never likely to attain individual historical interest or monetary value when compared with the best products of London makers and the masters of the craft. But they have a place in the story of our county. They witness to a skill akin to that which made Cornwall and its engineers a power in the world.

They have shared the hours of gladness and pain, the events of birth and death, the rise and fall of family fortunes. They have been inherited, bought and sold. They have served each owner with the same fidelity and impartiality, until they have become part of the lives of the Cornish people.

REFERENCES

1. Buller Records, County Record Office (CRO), No 360, 10 June 1689.

2. *Clock and Watch Makers in Wales*, Iorwerth C. Peate, Cardiff, 1960, p. 22.

3. Ibid, p. 24.

4. *Histories of Launceston and Dunheved*, R. & O. B. Peter, Plymouth, 1885, passim.

5. Henderson, MSS, vol 8, 1328 at the Royal Institution Library, Truro.

6. Agreement among the Johnstone papers from Trewithen, not yet catalogued, CRO.

7. *English Domestic Clocks*, Cescinsky & Webster, London, 1913.

8. Pet. 24, Royal Institution Library.

9. *Life of Richard Trevithick*, by Francis Trevithick, London, 1872, Vol. 1, p. 103.

10. *Memorials of Lostwithiel*, Truro, 1891, p. 77.

11. Will of John Pye, St Columb, proved 1740, with inventory.

The Wills referred to in the text and in the List of Makers will be found in the Archdeaconry of Cornwall Probate Records at the County Record Office, Truro.

GLOSSARY

Arbor: The horological term for axle or spindle.

Chapter-Ring: That part of the dial on which the hours are marked. Up to about 1790 (in Cornwall) it was usually a separate brass or silvered ring affixed to the dial plate.

Escapement: The device which regulates the movement of the mechanism, by permitting the wheel-train to move forward at exact intervals, and in return receives from the source of power (weight or spring) sufficient impulse to keep it in action. The more usual forms of escapement referred to in the text are chronologically:

i. The Verge escapement. The last wheel of the series, or train, has pointed teeth at right angles to the plane of the wheel. Across the wheel lies the verge or staff, bearing two projecting pieces at right angles. These are the pallets, and move in and out of the path of the teeth, which give an oscillating movement to the verge as each tooth escapes or passes. In the earliest examples the verge was vertical, surmounted by a balance wheel or by a foliot (qv). Later, after the invention of the pendulum as a controller of the mechanism, the verge was placed horizontally, and the escape or crown wheel arbor would be vertical and driven through a contrate wheel with teeth at right angles to the plane of the wheel but of normal shape to engage in the pinion. The verge escapement is the most primitive of all.

ii. The Anchor escapement, invented about 1671, has the teeth of the escape wheel in the same plane as the wheel. The pallets are somewhat like an anchor in shape, affixed to the horizontal arbor which carries the crutch to engage with the pendulum. The swing of the anchor moves the pallets in the path of the teeth, allowing one to escape with each complete swing.

iii. The Dead-beat escapement has pallets of a form which allows the teeth to fall dead at each swing, however great it may be, thus obviating the recoil or pushing back of the teeth, a fault

GLOSSARY

of the two previous escapements. This recoil interferes with the timekeeping property of the pendulum.

   iv. Cylinder escapement. In watches, a form in which the wheel-teeth fall upon a hollowed-out part of the staff or arbor of the balance wheel and pass through it as the balance swings.

Foliot: The first clocks were controlled by the inertia of a wheel or cross-bar attached to the vertical verge. The cross-bar is termed a foliot, and gave a slow swing, the period of which could be adjusted to some extent by shifting small weights hung on it towards or away from the centre.

Fusee: A conical-shaped pulley interposed between the mainspring and the wheel-train in a spring-driven clock to compensate for the different pulls when the spring is fully-wound or nearly run down.

Hammer-stop: The solid piece on which the striking hammer falls to be kept just clear of the bell, and so avoid an unpleasant jarring sound.

Royal Pendulum: The smaller swing of the anchor escapement allowed the use of longer pendulums, and a pendulum beating seconds, i.e. of about 39 in., or any longer pendulum, was referred to as a 'royal' pendulum on account of its superiority in timekeeping.

Spandrel: The square dial with round chapter-ring has four corner-spaces. These are termed spandrels and are often filled with castings or engravings of an ornamental sort.

Striking Mechanism: The sounding of the hours by a clock is usually accomplished in one of two ways:

   i. Count-Wheel or Locking-plate. The number of blows to be struck is controlled by a wheel with pins or slots set at increasing distances apart to allow the blows from one to twelve to be made. A lever falls into the slot to lock the mechanism until the next hour.

   ii. Rack-striking. The above form suffers from the disadvantage that the striking must be in succession, and any error is carried on until the clock is set right. In rack-striking a snail-shaped cam affixed to the hour arbor governs the number of teeth in a segmental rack of twelve teeth to fall past a given point, and as each blow is struck a 'gathering pallet' returns the rack one tooth to its resting-place.

LIST OF CORNISH CLOCKMAKERS

THE following list contains, under the heading of the towns of Cornwall, the makers and tradesmen found up to 1870, in alphabetical order. The period of activity of each has been indicated, as far as possible, from MS records, such as entries of baptism, marriage, or burial; from some printed notice, such as an advertisement for apprentices, or from similar sources. Failing these, a judgment of the date of some clock reported has been made; in these last instances the note 'c' will be found against the entry in the list.

Several names of later makers appear in commercial directories of the late 18th and early 19th centuries. These are scarce books, and it has seemed worthwhile to repeat the names which they contain, as the average reader is not likely to have ready access to them. Not a few clocks bear these names, and the entries will satisfy the curious, as well as shewing the long continuance in trade of some families and firms.

Those names of Cornish makers to be found in the comprehensive volume *Watchmakers and Clockmakers of the World*, Baillie, London, 1947, will be distinguished by the letters 'Ba'. In one or two cases the name or dates given in that book appear to shew a discrepancy with what local records disclose. In these few instances a note of Baillie's dates, etc., is appended without comment. Otherwise the particulars he gives are not repeated unless they add to our information.

It should be borne in mind that the entry of a name in the list does not imply that that person was actually engaged in business on his own. Journeymen, apprentices, repairers and the like may have found inclusion through reference in baptism, marriage or burial records to their trade as 'watchmaker' or 'clockmaker'. Unless some clock attributed to them is forthcoming, their precise status must remain largely obscure. And on occasion the name of an owner may appear on a dial, and not that of the maker, still further complicating matters; but these instances are probably rare.

Where a clock ascribed to a maker in the list has been examined or reported, brief mention of its features will be made. If more than one has been so examined, a typical specimen or specimens will be noted. The recording of a clock or clocks in the list may be taken as evidence that that clock exists at present, unless the contrary is stated. Many, if not most, are in private hands and not available, of course, for general inspection. With a very few exceptions all the clocks in the list have been personally examined by the author.

The terms 'Watchmaker' and 'Clockmaker' largely overlap, though in all probability little making of watches was done in the county. The descriptions of the trade in the list are as they exist in some written or printed record of the particular maker in his own day. Where there is no such description the general term 'clockmaker' may be assumed to apply. It may be well to remind the reader that there are likely to be few actually making clocks after about 1820, and that many recorded after that date were merely finishers or dealers, and never made a clock in their lives.

With the foregoing in view the list given should serve as a reasonably complete outline of Cornish clockmaking and those who practised it up to the middle of last century, and it is offered to the reader in the hope that it may prove useful to him in his enquiries.

ABBREVIATIONS USED IN THE LIST

Ba	Baillie, *Watchmakers and Clockmakers of the World*. London, 1947.
BD	Bailey's *Western, Midland and London Directory for the Year 1783*. Birmingham, 1783.
Bond	Signifies that the name appears in a Bond upon administration of goods.
c	Signifies a date, approximate only, estimated from details of a clock or clocks bearing this man's name. For 19th century clocks the dates can be little more than tentative suggestions.
CC	Member of the Clockmakers' Company.
CDP	Coulson's *Directory of Penzance and Neighbourhood*. Penzance, 1864.
EML	The Marriage Licences of the Diocese, preserved at Exeter.
KD	Kelly & Co, *Post Office Directory of Cornwall*. London, 1856.
KDD	Kelly & Co, *Post Office Directory of Cornwall*. London, 1873.
LC	Longcase or popularly, 'grandfather', clock.
OCW	Britten, *Old Clocks and Watches and their Makers*, ed. Baillie, Clutton and Ilbert, London, 1956; mostly in Ba.
pd	Pigot & Co, *Royal National and Commercial Directory and Topography*. London and Manchester, 1823.
PD	Pigot & Co, *Royal National and Commercial Directory and Topography*. London and Manchester, 1844.
RCG	The *Royal Cornwall Gazette* newspaper.
SM	The *Sherborne Mercury* newspaper.
UBD	The *Universal British Directory of Trade, Commerce and Manufacture*, 4 vols. London, 1791.
WB	The *West Briton* newspaper.
WD	Williams's *Commercial Directory of the Principal Market Towns in Cornwall*. Liverpool, 1847.
wp	Watchpaper, bearing advertising matter, etc., of this firm.

LIST OF MAKERS

The Names of Makers, and the Towns in which they worked, will be found in alphabetical order. In a few cases where there are several of the same surname, the order of Christian names will be chronological in order to show the relationships, and development of the business.

ALTARNUN

Cowling, C., 1856. Watch and clockmaker. KD.

Hender, William, 1848. Clock and watchmaker. Will 1848. See Callington.

ALVERTON, Penzance

Fox, John, 1843. Watchmaker. WB 22 Dec 1843, died at St Mary's, Isles of Scilly, wife of Mr Fox, watchmaker, aged 34. LC named on dial, but movement by J. P. Vibert.

BODMIN

Arnold, Richard. I, 1692/3. 'Pd Mr Richard Arnold for mending the clock and town beam, 4s 8d.' Mayor's a/cs, also 1695/6, 1699/1700. Will, 1724/5, son Richard (II) gets working shop, tools and implements. Witnesses include John Belling I, see under.

Arnold, Richard, II, 1737. Gunsmith and clockmaker. Fore Street. 'Arnold the gunsmith', in Lanhydrock estate a/cs, 1737. Mention in notebook of John Belling I. Leaves in will 1755 to elder son John (I) the sum of one shilling 'when he shall lawfully demand the same and not before'; gunsmith tools to son William and grandson John (II), 'share and share alike'. Daughters Sarah and Catherine, and widow Ursula to use the dwelling house 'in the fore street'. John Belling II a witness.

Arnold, John, I, 1754. Clockmaker. Elder son of above. See Launceston (same man?).

Arnold, John, II, 1736-1799. Eminent horologist. Improver of chronometer and inventor of refinements. Adelphi Buildings, London. Later established a factory at Chigwell, Essex. Plaque erected at Arnold's Passage, probable site of business premises of family in Bodmin.

Belling, John, I, 1685-c. 1761. Clockmaker. Fore Street. EML 4 Jun 1723 to Joan Comer. Corporation lease 1734 to John Belling, clockmaker, of stable and garden adjoining south Assize Hall. JB then aged 49, and son John (II) aged 11 in same lease. Ba. LC brass square dial 8-day dated 1706. Lantern clock, single hand, 30-hr. Bracket clock, brass arch dial, 8-day, quarter repeater on six bells. Turret clocks at Treworgey and Looe. Probable trade connection with the Arnolds.

Belling, John, II, 1753/4. Watchmaker. Fore Street. Married Armand Hoar 27 Jan 1753/4. SM 2 Apr 1770, JB jun. announces apprentice absconded. Will 1807. Leaves Rode Parks and tools, etc., to son John III. Property to daughters. A Thomas Ellicott witnesses will. UBD.

Belling, John, III, 1791. Watchmaker. Fore Street. Sons John (IV), William, James. named in will 1814. UBD.

Belling, John, IV, 1823. Watchmaker, ironmonger and insurance agent. Fore Street. WB 28 Jan 1848 selling off watches, clocks, jewellery, guns, cutlery, etc. pd, PD, WD, KD.

Belling, Miss Elizabeth, 1873. Watch and clockmaker. Fore Street. KDD.

Broad, Richard, 1785. Watchmaker. Fore Street. Mentioned in sale catalogue 1789. Bond 1825. UBD, Ba, pd. LC brass square dial, 30-hr, c. 1785.

Broad, John, 1809. Watchmaker and silversmith. Son of above. RCG 16 Dec 1809, boys confess plan to rob shop and take watch. Bond 1825. Ba, 1790-1820. PD. 1856, Honey Street, KD. LC painted arch dial, 30-hr, c. 1815.

Broad, Joseph, 1805. Clockmaker. RCG 9 Mar 1805, married. 31 Jan 1813, daughter baptised (parish Registers). Brother of above.

Broad, William Henry, 1873. Watchmaker. Honey Street. KDD.

Clement, ——?, 1695/6. 'Pd Mr Clement for mending the clock and chimes, £5', Mayor's a/cs.

Dunster, Matthew, 1724. Clockmaker. EML to Ann Betty of Exeter.

Ellory, Walter, 1647/8. 'Pd Mr Walter Ellory for and towards the mending of the clock, etc., 4s 3d.' Mayor's a/cs.

Gatty, Thomas, 1844. Watchmaker. St Nicholas Street. WB 27 Mar 1846, daughter dead. OCW gives dates 1790-1820. PD, KD, KDD.

Ham, Henry, c. 1810. LC painted arch dial, 8-day.

Hender, Edmund, 1847. Ironmonger and watchmaker. Fore Street. WD. LC square painted dial, 30-hr, c. 1820.

Henwood, Digory, c. 1770. LC brass arch dial, 8-day.

Hocke (Hawke), Christopher, 1661. Clockmaker. Married Eustes Blake at Bodmin. Repaired the tower clock at Lostwithiel, and made a 'Watch or Dyall upon the tower of the said Borough' for which he received the sum of £3 10s from the Mayor, Nevill Peeters. He contracted to keep the clock for seven years at 12d per annum (1671). Also agreed with Mayor of Grampound for a new clock for St Naun-

LIST OF MAKERS 65

ter's chapel, for which he received £6 (1673).

Jonson & Abraham, 1844. Watchmakers and silversmiths. Fore Street. WB 6 Sep 1844, sale on removing to Falmouth. PD.

Maunder, Michael, 1844. Watch and clockmaker. Fore Street. PD, WD, KD.

Pennington, Bernard, 1662/3. 'Pd the young Bernard Pennington for mending the clock, 5s.' Mayor's a/cs.

Reynolds, James, 1844. Watch and clockmaker. Also at Wadebridge. PD.

Scobell, John, 1645/6. 'Pd John Scobell for mending the clock, 5s.' Mayor's a/cs.

Signio (or Syneo), Richard, 1658/9. 'Pd Richard Signio for mending the clock, 6s 8d.' Mayor's a/cs. Inventory of tools, 1660, includes 'several tooles and other materials belonging to his trade, 15s, one little alarme belonging to a clock, 10s, one jack, 4s, two vises, one anvill and one pare of billows, 20s.'

Treleaven, James, 1847. Watch and clockmaker. Fore Street. WD.

Treleaven, Thomas, 1844. Watch and clockmaker. Fore Street. PD, KD, KDD.

Veale, ——?, End of 16th cent. Self-taught clockmaker, millwright, etc.

BRADOC (Broad Oak)

Smith, Owen, 1784. Ba—'left 1784'. Possibly attended the clocks at Boconnoc.

BREAGE

Tregoning, ——?, 1770. Rev William Borlase's a/c book—'to Tregoning of Breag for fitting the kitchen clock'.

Tyacke, George, c. 1820. Will 1853. LC brass arch dial, plain, 8-day.

CALLINGTON

Blight, Edward, 1873. Watchmaker. KDD.

Budge, W., 1856. Watch and clockmaker. WB 17 Oct 1856, daughter born. KD.

Budge, Mrs Elizabeth, 1873. Watchmaker. KDD.

Hay, George, 1785. EML. LC brass dial, 30-hr, c. 1800.

Hender, William, 1814. Ironmonger, clock and watchmaker. Fore Street. RCG 19 Nov 1814, 3 silver watches stolen. WB 23 Mar 1832, declining business. Died 1848. pd, PD. LC painted arch dial, Arabic numerals, 30-hr, c. 1820.

Hoskin, Richard, 1844. Watch and clockmaker. Higher Street. PD, WD.

Philp, James, 1844. Watchmaker. Higher Street. PD, WD.

Treweek, John, 1823. Watchmaker. Fore Street. pd.

Wadge, Josiah, 1754. Clockmaker. Bond 1754 on Honour Wadge of Liskeard. SM 18 Nov 1754 debts to be declared to Josiah Wadge of Callington, clockmaker.

Wadge, Agrippa, 1791. Watch and clockmaker. UBD, Ba.

Wadge, R. Dodge, 1791. Watch and clockmaker. UBD, Ba.

Wadge, Josiah, 1823. Watchmaker. Lower Street. pd.

Wadge, Josiah & Francis, 1844. Watchmakers. Lower Street. WB 25 Jan 1856, Francis dead, aged 62. PD, WD. J. Wadge, 1856, KD.

Williams, John Thomas, 1856. Watch and clockmaker. KD, KDD.

CAMBORNE

Anear, Henry, 1873. Watchmaker. Church Lane. KDD.

Gwilleam, John, 1787. Watchmaker. Married at Falmouth to Mary Benalack.

Hockin, John, I, 1791. Clockmaker. Married Mary Champion. Will 1805. Residuary legatee, son. LC brass arch dial, 8-day, dated 1805.

Hockin, John, II, 1805. Clockmaker. 'Lately come to reside.' pd.

Hocking, Samuel, 1856. Watch and clockmaker. Market Place. KD, KDD.

Hocking, William, 1873. Watchmaker. Tuckingmill. KDD.

Hosking, John, 1844. Watchmaker. Market Street. WB 7 Oct 1853, dead. PD, WD.

Menadue, William, 1844. Watch and clockmaker. PD. See also St Agnes.

Nicholl, Thomas, 1873. Watchmaker. Trelowarren Street. KDD.

Nicholls, Joseph Henry Penrose, 1873. Watchmaker. Tuckingmill. KDD.

Pollard, William, 1844. Watchmaker. Church Row. WB 10 May 1848, daughter born. PD, WD. Market Place, 1856, KD.

Pollard & Son, 1873. Watchmakers. Market Place. KDD.

CAMELFORD

Clode, William, c. 1760. LC brass square dial, 30-hr, wheels within plates as 8-day. Formerly hour-hand only. Motion work for minute-hand added later.

LIST OF MAKERS

Ede, John, 1844. Watch and clockmaker. Fore Street. PD, WD, KD, KDD.
Hender, Nathanael, c. 1760. LC brass dial, 30-hr.
Hocken, Henry, 1823. Watchmaker. pd.
Hocken, John, 1856. Watch and clockmaker. Fore Street. KD, KDD.
Hocken, William, 1847. Watch and clockmaker. WD.
Hosking, Henry, 1814. Watchmaker. RCG 31 Dec 1814, married to Miss D. Carew at Camelford.
Pascoe, Charles, 1847. Watch and clockmaker. WD.
Peter, John, 1832. Watchmaker. WB 6 Apr 1832, emigrated to America.
Slade, ——?, 1791. Watch and clockmaker. UBD, Ba.

CHACEWATER

Anear, C., 1856. Watch and clockmaker. KD.
Anear, Henry, 1852. Watchmaker. WB 13 Feb 1852, son born.
Hocking, James, c. 1830. Watch and clockmaker. PD. LC painted arch dial, 8-day.
Tonkin, James, ? LC painted arch dial.

CONSTANTINE

Bolitho, S., ? LC painted dial, 8-day.

CORNELLY (see Tregony)

EGLOSHAYLE

Champion, Thomas, 1831. Watchmaker. Son, Thomas Taylor, baptised 8 Apr 1831 (parish Registers).
Reynolds, John, c. 1780. LC brass square dial, 8-day, name of town spelt 'Egleslail'. Another, brass square dial, 8-day, with name and 'Wadebridge' by same maker. A John Reynolds buried at Egloshayle in 1786.

FALMOUTH

Baker, Thomas, c. 1810. LC painted arch dial.
Beringer, Joseph & Sons, 1873. Watch and clockmakers. 12 Market Street. KDD.
Beringer & Schwerer, 1847. Watch and clockmakers and outfitters. Market Street. WD.

Carlyon, John, 1729. Clockmaker. EML to Ann Jenkins of Falmouth. Bond 1753. LC brass arch dial, 8-day, c. 1740. Also sundial.

Carver, Alfred, 1873. Watchmaker and jeweller. 32 Market Street. KDD.

Cumming, James, 1783. Watchmaker and silversmith. BD, Ba. LC brass arch dial now silvered. Windmill, two sawyers and grindstone in arch.

Davidson, John, 1847. Watch and clockmaker. Ludgate Hill. WD.

Dunstan, James, 1783. Silversmith and watchmaker. BD, Ba.

Eva, Richard, 1734-1806. Watch and clockmaker and silversmith. From Tregony about 1780. Married 1782 to Margaret McDowell. Patentee 1796. UBD, Ba. Good Bracket clock, c. 1785. LC with planetarium in arch (OCW). See Tregony.

Falkson, Lewis, 1787-1852. Watch and clockmaker. High Street, PD. Ludgate Hill, WD.

Genn, James Ditchett, 1844. Watch and clockmaker. Market Street. PD, WD, KD.

Goffe, William, c.-1800. Watchmaker and chronometer maker to His Majesty's Packets. The Moor. LC painted arch dial, 8-day, strike/silent in arch, brass-cased weights, superior. Pocket chronometers with his name 1825, with Earnshaw's spring detent, steel helical balance spring.

Gulliam, William, 1791. Watch and clockmaker. UBD, Ba.

Hambly, William, 1783. Watchmaker, silversmith and clockmaker. BD, UBD, Ba.

Harris, Samuel, 1823. Watchmaker. High Street. pd.

Henry, ——?, Abt 1790. CC, Ba.

Howells, William, 1790. CC. RCG 28 Mar 1812, Chronometers, watches and clocks. Appointed by Their Lordships H.M. Postmasters General Chronometrician to H.M. Packets, Falmouth. A 'transit room will shortly be built with an instrument by Troughton which will enable Mr Howells to give the most correct rates and time'.

Jacob, Moses Jacob, 1813-1860. Watch and clockmaker, dealer and pawnbroker. Arwenack Street. WB 15 Dec 1854, offers watches. President, Falmouth Hebrew congregation. KD.

Jacobs, S(olomon?), c. 1800. Bracket clock, 8-day, enamel dial, anchor escapement, lancet-style mahogany case. Possibly as under Penzance.

Johns, John William, 1856. Watch and clockmaker. Market Street, KD. 4 Church Street, KDD.

Joseph, Abraham, 1827. Watchmaker. Market Street. WB 3 Aug 1827, son leaving town, partnership dissolved.

Joseph, Joseph, 1823. Watchmaker. Market Strand. pd.

Kendal, Edward, 1753. Watchmaker. Bond. 1753.

Martyn, John Nicholls, c. 1800. Watchmaker. Church Street. Ba, pd. Fine mahogany-cased regulator with centre-seconds hand, 8-day, round silvered dial.

Merifield & Genn, 1823. Watchmakers. Market Street. pd, Ba.

Nancolas, Anthony, 1783. Silversmith and watchmaker. 1784, married Elizabeth Warne (parish Registers). BD. LC brass arch dial, 8-day, face in arch with moving eyes, good mahogany case, c. 1785.

Olive, John, 1844. Watch and clockmaker. Church Street. PD.

Olive, Michael, 1849. Chronometer, watch, clock and nautical instrument maker. Church Street. WB 22 Aug 1856, advertises for apprentice. KD. 26 Arwenack Street, KDD.

Olive, Thomas, 1849. Chronometer maker. Church Street. WB 24 Aug 1849, advertises sale on death, including astronomical clock purchased by brother (above) who carries on business.

Rowe, Benjamin Chalwenn & J., 1823. Watch and clockmakers. High Street. pd, PD. Ludgate Hill, WD.

Schwerer, Jacob, 1844. Watch and clockmaker. Market Street. PD.

Smith, Charles, 1813. Watchmaker and jeweller. Market Strand. RCG 5 Jun 1813, advertises sale of stock. pd.

Webber, John, 1844. Watch and clockmaker. Market Street. PD, WD, KD, KDD.

Webber, William, 1845. Watchmaker. WB 11 Jul 1845, daughter born.

Wheatley, W., jun, 1856. Watch and clockmaker. Market Street. KD.

Williams, Arthur, 1814. Watchmaker. Church Street. WB 15 Jul 1814, son born. pd, PD, WD, KD.

Winterhalder, Matthew, 1873. Watchmaker and jeweller. 20 Church Street. KDD.

FOWEY

Acland, Henry, 1844. Watch and clockmaker. PD.

Bennet, J., 1856. Watch and clockmaker. KD.

Henwood, Digory, 1765. Clockmaker. Daughter born (parish Registers). See Bodmin.

Hewett, Richard, 1791. Ironmonger, silversmith and watchmaker. SM 25 Jun 1792, burglary at shop. WB 22 May 1846, wife dead. 20 Nov 1846, sale of business by Miss Hewett. Will 1847. UBD, Ba, pd, PD.

LC brass arch dial, plain, 8-day, c. 1790.

Hodge, William, 1873. Watchmaker. KDD.

Orchard, Henry, 1844. Watch and clockmaker. PD, WD.

Rawe (or Rowe), James, 1778. Watchmaker. Married Mary Little. Will 1781. SM 3 Apr 1797, watch with this name given to Mr Sholl of Truro.

Varco, Robert, jun., 1873. Watchmaker. KDD.

Vivian, James, 1873. Watchmaker. KDD.

GRAMPOUND

Clemmow, C., 1856. Watch and clockmaker. KD.

Thoms, William, 1663. 'For one quarter for his attendance, kepeing of the clock, and Ringing of ye Bell, 15s.' Mayor's a/cs.

Tregosse, Stephen, 1686. 'For kepeing the clock, 2s 6d.' Mayor's a/cs.

GUNNISLAKE

Long, R., 1856. Watch and clockmaker. KD.

HAYLE

Coad, James Thomas, 1873. Watch and clockmaker. Penpol Terrace. KDD.

Harvey, Francis, 1786. Silver pair-case verge watch, dated 1786.

Ingram, John, 1844. Watch and clockmaker. WB 7 Jul 1848, daughter born. PD, KD.

Ley, Thomas Stephen, 1844. Watch and clockmaker. PD.

Ley, William, 1847. Watch and clockmaker. Copperhouse. WD.

Phillips, John, c. 1820. LC painted arch dial, moon-phase in arch, 8-day.

Rice, Nicholas, 1844. Watch and clockmaker. PD. See also Penryn.

Wearn, Roger, II, 1748-1820. LC brass dial. See St Erth.

HELSTON

Bennett, John, 1775. Watchmaker. SM 29 Mar 1779, bankrupt. 19 Apr sale of stock, includes smoke and other jacks, turret clocks complete. Ba. LC silvered arch dial, engraved centre, 8-day, white and gold Adam case (spoiled), c. 1775.

Bennett, Samuel, 1763. Clockmaker. '22 Sep 1763, paid Bennett for mending my watch.' Trewinnard ledger, Trewithen MSS. 1768, married Catherine Pearce (parish Registers).

LIST OF MAKERS

Beringer, John, 1847. Watch and clockmaker. Meneage Street. WB 4 Jan 1850, daughter born. WD, KD.

Beringer, Joseph, 1844. Watch and clockmaker. Meneage Street. PD.

Beringer, Joseph & Sons, 1873. Watchmakers and jewellers. Meneage Street. KDD.

Coulson, William, c. 1820. LC painted arch dial, 8-day.

Curgenven, Henry, 1763. EML. LC brass arch dial with 'Success to the British Fleet', 8-day, c. 1785.

Dobb, Henry, 1809-1852. Watch and clockmaker. Coinagehall Street. WB 2 Jul 1852, dead, aged 43. PD, WD.

Hodge, Jonathan, 1844. Ironmonger, watch and clockmaker, etc. Meneage Street. Had as apprentice Hodgson Pascoe (qv under Penzance). PD, WD, KD. Bankrupt 1860.

Julian, Thomas, 1873. Watchmaker. Meneage Street. KDD.

Lane, Charles, 1823. Watch and clockmaker. Coinagehall Street. WB 13 Jan 1854, wife dead. pd, PD, WD, KD.

Lane, William Puckey, 1873. Watchmaker and jeweller. Meneage Street. KDD.

Martin, Richard, c. 1780. LC brass square dial, oak case, 8-day.

Odger, Thomas, c. 1800. Watchmaker. Coinagehall Street. Bond 1823. LC regulator, round dial and hood, 8-day, c. 1800. pd.

Polglaze, Henry, 1734. Clockmaker. Bond 1734.

Read, Charles, 1788. Watch and clockmaker and mercer. Married 27 Jan 1788 Mary Mabyn (parish Registers). SM 14 Mar 1803, robbery—house plundered, but not shop.

Read, John Mabyn, c. 1815. Watch and clockmaker. Meneage Street. Son of above. pd, PD. LC brass arch dial, 8-day. wp dated 1818.

Shepard, Thomas, ——? Watch reported, no details.

Symons, W. B., 1856. Watch and clockmaker. Meneage Street. KD.

Thomas, Richard, I, 1781. CC. Watchmaker. Married Joan Richards 1781 at Breage. Ba.

Thomas, Richard, II, 1804. Watchmaker. Married Jenefer Polamountain 8 Aug 1804 at Helston. RCG 28 Apr 1804, watch lost; maker Richard Thomas, Helston, No 2954. LC painted arch dial, Arabic numerals, c. 1815.

Troughton, John, 1707/8. 'Pd John Troughton for keeping the clock.' (Penryn Town a/c book). 'Pd Mr Troughton for mending N. Tresadern's watch twice, 6s.' 1723. East Pool Mine Cost Books, Trewithen MSS. Kept clocks of Rev William Borlase, of Ludgvan, between 1735 and 1743. Fine LC brass arch dial, oak case, 8-day, c. 1750.

West, William, 1751-1831. See St Ives. Good LC brass arch dial with 'William West, Helstone'. Now in America. c. 1775.

KILKHAMPTON

Mill, John, 1856. Watch and clockmaker. KD, KDD.

Mules, William, c. 1830. Watch and clockmaker. LC painted square dial, 30-hr. See also Stratton.

KINGSAND

Ham, John, 1844. Watchmaker. PD.

LADOCK

Randell, William, 1647. Watchmaker, gentleman. Epitaph in parish church.

LAUNCESTON

Arnold, John, 1754. Clockmaker. SM 28 Oct 1754, news of lost watch to John Arnold, clockmaker, Launceston. See Bodmin.

Box, John, 1823. Ironmonger, watch and clockmaker. Westgate Street, pd. Broad Street, PD, WD.

Braund, Thomas, 1844. Watchmaker. Bond 1844.

Coryndon, William, 1752. EML. LC brass square dial, 30-hr, c. 1790.

Dingley, Richard, 1823. Watchmaker. Back Street. pd.

Forrest, Jonathan, 1757. EML. SM 5 Dec 1757, watch stolen from shop. SM 9 Jan 1758, lost, double gold case watch to be returned. Tavern clock, round head, short trunk, now 8-day but altered, c. 1780.

Harvey, Emanuel, 1775. Watch and clockmaker and shopkeeper. Married 26 Sep 1775. Will 1818, to son Edward, all clocks, watches, plate, tools and implements of trade. UBD.

Honey, Thomas, 1884. Watch and clockmaker. Broad Street. PD.

Maunder, Aaron, 1856. Watch and clockmaker. Church Street. KD, KDD.

Nattle, Richard, 1791. Watch and clockmaker. UBD, Ba.

Nicolls, Henry, 1791. Watch and clockmaker. LC painted arch dial with 'Henry Nicolls, Launceston', 8-day. See St Austell.

Pearce, Edward, 1844. Watch and clockmaker. Madford Lane, PD; Race Hill, WD; Southgate, KDD.

LIST OF MAKERS

Pendray, Digory, 1734. Clockmaker and gunsmith. Married 1734. Inventory 1741, total £62 1s 5d includes: 3 dozen of new small clock files 1s 8d; 3 dozen and 8 new small clock files of different sort 8s 0d; 8 new springs for watches 8s 0d; 4 dozen ½ of keys for watches and 3 dozen of chrystals 7s 6d. In the Forge Shop: a large clock not finished £2 2s 0d; 3 new clocks and 2 ditto not finished £7 10s 0d; 1 old clock 7s 6d; 22 clock hands 1s 0d; 2 steel watch springs 2s 0d; three clock cases £1 1s 0d; a new clock on tryal sent to Benjamin Took of Marhamchurch £2 2s 0d; due for keeping the town clocks £2 0s 0d.

Proctor, James, 1801. EML. LC brass dial, 8-day.

Reynolds, Henry, 1873. Watchmaker. Westgate Street. KDD.

Reynolds, William, 1823. Watch and clockmaker. Westgate Street. WB 25 Mar 1853, son dead. pd. Church Street, PD. WD, KD. LC brass arch dial, moon-phase, 30-hr, c. 1820.

Reynolds, William, jun, 1856. Watch and clockmaker. High Street. KD.

Robyn, Walter, 1460. Clockmaker. For setting 'le clokke' with the time, Borough a/cs.

Routleigh, George, 1745-1802. (Sometimes Routledge or Roughleigh.) Watchmaker. UBD, Ba. Mostly LC brass square dial 30-hr. Epitaph in Lydford Churchyard, Devon. (See front of book.)

Scown, John, 1847. Watch and clockmaker and coffee-house. Westgate Street. WD.

Spry, Nathanael, 1823. Watchmaker. Southgate Street. pd.

Symons, John, 1847. Watch and clockmaker and hardware shop. Southgate Street, WD. Church Street, KD. KDD.

Uglow, Abel, 1823. Watchmaker. St Thomas Street. pd.

Upjohn, Thomas, 1765. EML. Possibly the Devon maker of this name. Ba.

Wilkins, John, 1735. Clockmaker. Married 1735.

LISKEARD

Abraham, Josiah, 1796-1879. Watch and clockmaker. Used movements made by the Hams (see below). Church Street, PD. Pike Street, WD, KD. Castle Street, KDD. Silver pair-case verge watch, 1832. LC silvered brass arch dial, 8-day, c. 1840.

Austen, Richard, 1823. Watchmaker. Parade. WB 8 Aug 1845, dead. pd. Church Street, PD.

Austen, Thomas, c. 1760. LC brass 10 in. square dial, engraved centre,

30-hr. Movement has turned corner pillars, as lantern clock. The Austens made their own clocks, the patterns being drawn on parchment. Grandfather of above.

Botterell, James Hosking, 1873. Watchmaker. Fore Street. KDD.

Broad, William, 1873. Watch and clockmaker. Middle Lux Street. KDD.

Clarke, ——?, 1831. Watchmaker. WB 9 Sep 1831, robbery on shop.

Davey, John, 1847. Watch and clockmaker and Registrar. Church Street. WD.

Foot, John, 1769. Clockmaker. SM 18 Dec 1769, news of lost watch to JF. 'Pd John Foot for a new key for the clock case,' Buller MSS, CRO. 1784-6, 'John Foot's bill for the clock,' Mayor's a/cs. LC brass square dial engraved centre, 30-hr.

Geach, George, 1811. Clockmaker. EML. WB 13 Mar 1818, shop lighted with gas.

Ham, Joseph, c. 1775. Mayor's a/cs keeping clock, 1805-6. Chiefly LC brass dial 30-hr, also supplying local dealers. Tools used by the Hams remain.

Ham, John, c. 1795. Son of above. Moved to Kingsand and Millbrook. LC painted arch dial, 8-day, c. 1800.

Ham, William Pote, c. 1820. Watchmaker. Fore Street. KDD. LC painted arch dial, 8-day, moon-phase, c. 1820.

Hitt, Peter, 1748. Clockmaker. 'Peter Hitt's bill for cleaning the clock,' Fortescue Estate Papers, CRO. 1777-84, 'Mr Hitt's bill for repairing the clock,' Mayor's a/cs. Bond to Mary his widow 1803. LC brass arch dial, 8-day with 'High Water at Plymouth Key' in arch.

Hodge, John, c. 1800. LC brass square dial, engraved centre, 30-hr.

Kaupp, Francis Joseph, 1873. Watchmaker. Higher Lux Street. KDD. wp's 1877-1886. Retailer only.

Langlye, William, 1605. 'Pd William Langlye for making a clock upon the Market Hall, £2 18s 4d.' Mayor's a/cs.

Libby (or Lebby), Daniel, 1791. Watchmaker. UBD. LC brass dial 8-day, c. 1790.

Mayell, Edwin, 1844. Watch and clockmaker. Tavern Street. PD, WD, KD.

Moon, R., 1856. Watch and clockmaker. Tavern Hill. KD.

Murray, William, 1823. Watch and clockmaker. Church Street. Died 1883. pd, PD.

Mutton, Samuel, 1847. Watch and clockmaker. Church Street. WD.

Pascoe, Charles, 1856. Watch and clockmaker. Fore Street.

LIST OF MAKERS 75

Retallick, Richard, 1810. Watchmaker. Market Street. Mentioned in will 1810. pd. LC painted dial, 30-hr, movement probably by the Ham firm.

Roger, William, 1650. 'Pd W.R. the younger for mending the clocke 18s 0d', Mayor's a/cs, 1650.

Roskilly, William, 1856. Watch and clockmaker. Fore Street, KD. Lower Lux Street, KDD.

Row, F., 1850. Witness to Will, 1850.

Sampson, Richard, 1734-1814. Watchmaker. Formerly of St Columb. WB 18 Mar 1814, died aged 80. LC brass arch dial once 30-hr now 8-day, c. 1790.

Sampson, Robert, 1791. Watch and clockmaker. UBD, Ba. Probably meant for the above.

Sleep, John, 1844. Watchmaker. Fore Street, PD. Market Street, WD. WB 16 Apr 1847, watch lost, return to Mr Sleep. LC painted dial 30-hr, c. 1840.

Smith, Charles, 1803. CC. Watchmaker. RCG 6 Aug 1803, married at St Veep. WB 16 Aug 1816, business to be disposed of. Particulars from John Smith. Ba.

Spry, Jacob, 1634. 'Pd Jacob Sprye for mendinge the clocke 3s', Mayor's a/cs.

Vian, William, 1729. Watchmaker. EML. 'Pd William Vian watchmaker 8s 6d', Trelawny a/cs, Jan 1739/40. Mayor's a/cs 1741-1762-76. LC brass square dial, 30-hr, single hand, c. 1740.

Volk & Floessel, 1873. Watch and clockmakers, jewellers, etc. Fore Street. KDD.

Webb, John, 1811. Watchmaker. Church Street, PD. Lower Lux Street, WD. RCG 25 May 1811, business to be let. Mr John Webb about to decline business. WB 30 Jun 1815, shop broken open, about 70 watches stolen.

Webb, W. H., 1856. Watch and clockmaker. Lower Lux Street. KD.

Webb & Abraham, 1823. Watchmakers. Lux Street. pd.

Williams, Joseph, 1873. Watchmaker. Fore Street. KDD.

Williams, Richard, 1774. SM 5 Dec 1774 Richard Williams of Liskeard wants apprentice to watch and clockmaking. LC brass dial 30-hr.

LOOE

Dinzell, John, 1728. 'Pd John Dinzell for looking after the clock and repairing it, £2 2s 2d' CRO, DDX 155.

Hender, Thomas Row, 1844. Watch and clockmaker. PD.

Martyn, J., c. 1780. LC brass engraved dial.

Powne, Thomas, 1806. Ironmonger and watchmaker. EML. WB 9 Sep 1814, 'supposed to be stolen from the cabin of the barge *Speculation*, of Looe, on Monday last about noon, at Mevagissey, a silver watch, maker's name T. Powne, No 6,389, inside case engraved Wm. Pengelley. Whoever will bring the same to Mr Powne, watchmaker, East Looe, shall receive £1 reward'. PD, WD, KD. LC painted arch dial, 8-day, c. 1815.

Wynhall, Edwin, 1856. Watch and clockmaker. KD, KDD.

LOSTWITHIEL

Abrahams, Aaron, 1821. Watchmaker. WB 16 Mar 1821, 'late of Lostwithiel, removed to 23 Nut Street, Plymouth'.

Belling, John, 1823. Watchmaker. Fore Street. pd.

Bennett, John, 1873. Watch and clockmaker. Fore Street. KDD.

Branwell (Robert Matthews?), c. 1798. Watchmaker. LC painted dial, 30-hr. Name very indistinct. See Penzance.

Broad, William, 1844. Watch and clockmaker. Queen Street. WB 19 Jan 1849, wife dead. PD, WD, KD.

Hawken, Thomas Edward, 1873. Watchmaker and photographer. Queen Street. KDD.

Hocking, John, 1824. Watchmaker. Daughter baptised at St Columb. Interesting letter, 8 Oct 1824—

Dear Father & Mother,

You sent me a letter to get that Eight-day clock finished it is impossible for me to finish a clock and do my own work I wood give Wm. two or three clocks to make for me if he conveniontly do it I may so only pay him as the Bristol workmen I ave moor work thin I can fitely do I can't keep a clock made in the shop I sent you a thirty our clock and mist the sale of one by the means of it, it is oute of my power so to do as I have toock a house of Mr Lanyon the rent and outes alltogether is £21-0-0 per year money enuf fo me to get. I remain your dutiful son J. Hocking.

LC painted arch dial, 30-hr, Arabic figures, c. 1820. Town name 'Lostwithirell'. (See page 44.)

Mayell, Joseph, 1823. Watch and clockmaker. Fore Street. WB 25 Jan 1828, journeyman wanted. pd.

Mutton, Samuel, 1823. Watchmaker. North Street. pd. (See Liskeard.)

LIST OF MAKERS

MADRON

Arthur, P., 1856. Watch and clockmaker. Hea Moor. KD.

MARAZION

Daniel, F. H., 1845. Watchmaker. WB 15 Aug 1845, daughter dead.
Daniel, Richard Thomas, 1823. Watch and clockmaker. pd, PD.
Ivey, William, 1873. Watch and clockmaker. KDD.
Sellick, James, c. 1740. LC brass square dial 30-hr single hand, engraved centre with scene.

MARHAMCHURCH

Box, William, c. 1830. LC painted dial, 30-hr.

MEVAGISSEY

Bell, ——?, c. 1770. Watch reported, no details.
Brewer, ——?, 1712. 'Pd Brewer of Mevagizzy for righting my best clock, 1s 6d', Tremayne Diaries, CRO.
Kimberley, Thomas, 1856. Watchmaker. KD, KDD.
Michael, Benjamin, 1768-1853. Watchmaker. See St Austell. WB 19 Aug 1853, dead suddenly 'whilst on his road to Mevagissey, aged 85 years, one of the oldest inhabitants of the town'.
Michell, Benjamin, 1785. Watchmaker. Married Rachel Bawden at Lostwithiel 12 Aug 1785. Bond 1823, to son Benjamin, tailor. LC brass arch dial, moon-phase, c. 1800.
Mitchell, John, 1819. Watchmaker. Bond to father, Benjamin, tailor, 10 Mar, 1819.
Slade, Edward, 1844. Watch and clockmaker. PD, WD. LC painted arch dial, 8-day, c. 1830.
Smith, Peter, 1789-1844. Watch and clockmaker. EML 1815. WB 19 Apr 1844, dead.
Truscott, Lewis, 1812. EML. Ba. LC brass square dial, 30-hr, c. 1800.
Veal, Joseph, c. 1790. LC brass arch engraved dial, 8-day.

MILLBROOK

Bennett, Robert, 1856. Watch and clockmaker. KD, KDD.
Ham, John, c. 1820. PD. LC painted arch dial, 8-day. Also at Kingsand.

MORWENSTOW

Harris, Aaron, c. 1830. LC painted dial, Arabic numerals, 30-hr.

NEWLYN (Penzance)

Kelynack, Richard, 1805. Dated silver pair-case verge watch.

NEWQUAY (St Columb Minor)

Reynolds, John, 1856. Watch and clockmaker. WB 20 Jun 1851, son born. KD, KDD.

PADSTOW

Best, John, 1772. Married Mary Martin, 7 May, 1772. LC brass square dial, 8-day, chip-carved case, probably later, c. 1780.

Boney, Caleb I, 1747-1826. Clockmaker. Will 1827, all goods, tools, and stock-in-trade to elder son and executor, John Boney. All Musical boxes and musical instruments and a bright strake used in block tin work to younger son, Caleb. To daughters of Caleb the wearing apparell of late wife Christian. UBD, Ba. Astronomical and musical clocks.

Boney, Caleb II, 1828. Ironmonger, tallow-chandler, clockmaker and victualler. Younger son of above. Insolvent, WB 24 Oct 1828.

Boney, John, 1844. Watch and clockmaker. Elder son of Caleb Boney I. PD, WD. Duke Street, KD.

Geach, George, 1823. Watchmaker. Market Place. pd.

Reynolds, John, 1820. Watch and clockmaker and photographer. Lanadwell Street. KD, KDD. LC painted arch dial with rocking ship, 8-day.

Reynolds, William, 1844. Watch and clockmaker. PD, WD. South Quay, KD. LC silvered brass arch dial, mahogany case, 8-day, c. 1840.

PENRYN

Behenna, Richard, c. 1810. LC brass plain arch dial with rocking ship, 8-day, c. 1810.

Cock, Richard, c. 1720. Bracket clock, 8-day, cherub spandrels arch dial with silvered boss. Back-plate engraved Richard Cock, Penryn, Fecit. c. 1720. 1728/9, Richard Cock, keeping clock, Penryn Town a/c book.

Cock, William, 1745. Clockmaker. EML to Elizabeth Trood, Penryn.

Harry, John, 1783. Jeweller, watchmaker and hardwareman. BD.

Hornblower, Jonathan? c. 1780. LC painted arch dial with brass spandrels and name plaque, 8-day. 'Time how short. Eternity how long!' in arch with figure of Time.

Olive, Frederick, 1844. Watch and clockmaker. Higher Market Street, PD, KD. Broad Street, KDD.

LIST OF MAKERS 79

Olive, Michael, 1847. Watch and clockmaker. Market Street. WD. See Falmouth.

Olive, Thomas, 1768. Watchmaker, silversmith and gunsmith. Lower Street. SM 21 Nov 1768, wants apprentice. 17 Jun 1782, burglary from shop. 20 Feb 1792, wants apprentice, 'he by steady practice and several years experience in London has attained a perfect knowledge in the trade'. Lessee of Lukey's Garden, Bohill, Penryn, 1769, 1775. BD, UBD, Ba, pd.

Rice, Nicholas, 1813. Watch and clockmaker. RCG 13 Mar 1813, apprentice wanted.

Slade, John, 1844. Watch and clockmaker. High Street, PD. Lower Market Street, KD.

Slade, William, 1844. Watch and clockmaker. Market Street. PD, WD. LC painted arch dial, 8-day, c. 1840.

Thomas, John Marshall, 1873. Watchmaker and jeweller. Higher Market Street. KDD. Also at Redruth.

Trenerry, John, 1652. 'Pd (John) Trenerry for mending the clock, 5s.' Town a/c book.

Wimpen, David, 1736. Clockmaker. EML to Elizabeth Gluvias of Falmouth.

PENZANCE

Arthur, Peter, 1844. Watch and clockmaker. Died 1857. East Street, PD. Market Jew Street, WD. LC painted dial, alarm, 8-day.

Beckerlegge, John, 1864. Watchmaker. 69 Market Jew Street. CDP.

Beringer, Fidelis, 1864. Watchmaker. 27 Market Place. CDP.

Beringer, John, 1844. Watch and clockmaker. Alverton Street, PD. St Clare Street, WD. Afterwards—

Beringer & Schwerer, 1856. Watch and clockmakers. Market Place, KD. 27 Market Place, CDP. 16 Market Jew Street, KDD.

Boney, Caleb I. See Padstow. A branch shop here c. 1800? LC brass arch dial, musical, 8-day. LC painted arch dial, 8-day, c. 1820.

Branwell, Robert Matthews, 1798. Watchmaker. Leased Vellan Hoggan Mills in Gulval 1807.

Carne, John, 1752. Watchmaker. Rev William Borlase's a/c book, 'Pd Mr Carne watchmaker for looking after the clocks to Xtmas 1752, 7s'. EML 1775. Ba.

Cock, Richard, c. 1720. SM 22 Apr 1771, watch with this name to be returned to John Carlyon, watchmaker, Penzance. See Penryn.

Coulson, Henry, 1772-1834. CC. Watchmaker. Apprenticed to John Sampson of Penzance. Own business 12 Market Place. Turned to drapery and hosiery, at which he made a large fortune. UBD, Ba. LC painted arch dial, 8-day, poor, c. 1815.

Daniel, Michael, 1847. Watch and clockmaker. Market Jew Street. WD.

Daniel, Nicholas, 1844. Watch and clockmaker. Chapel Street. PD, wp. LC painted arch dial, Arabic numerals, 8-day, c. 1840.

Davey, William, 1791. Watchmaker. UBD, Ba.

Harris, Morris Hart, 1844. Dealer, etc. Chapel Street. PD. Advertises American clocks at 17/6 each, 1847. Subscribed 2/6 for starving Jews in Tiberias, 1849. LC painted arch dial, 8-day, c. 1820.

Harvey, Richard Andrew, 1823. Watchmaker and ironmonger. Market Place. WB 24 Jan 1834, apprentice wanted. RCG 28 Mar 1835, bankrupt. pd.

Harvey, Thomas, 1840. Watch and clockmaker. Market Place. WB 21 Sep 1849, disposal of business, owner about to leave for India. LC silvered brass arch dial, engraved motifs in spandrels, 8-day, c. 1840.

Hosking, John, c. 1747. Brass square dial, 30-hr, posted frame, c. 1760. A John Hoskin (or Hosken) paid 8 gns. for mending town clock, 1747.

Hotten, John Thomas, 1852. Watchmaker and silversmith, dealer and chapman. WB 18 Jun 1852, bankrupt.

Jacobs, Levi, 1823. Watchmaker. WB 1 Aug 1823, stolen watch offered to him.

Jacobs, Solomon, 1803. Watchmaker. 11 May, Bond, Rebecca his widow. See also Falmouth. Ba.

Kistler, George, 1856. Watch and clockmaker. North Street. KD.

Kistler, Matthew, 1844. Watch and clockmaker. North Street. PD, WD.

Kistler, George & Matthias, 1873. Watchmakers. 6 Causewayhead. KDD.

Lawrence, Thomas Vigurs, 1811. Watchmaker. WB 12 Jan 1811, married at Madron after a courtship of eight days to Miss Sarah Ford, with a fortune of £700 per annum. 12 Jun 1817, suspected of theft.

Levin, Alexander, 1847. Watch and clockmaker and marine stores. Market Jew Street. WD.

Michell, Joseph, 1844. Watch and clockmaker. Parade Street. Died 1 Nov 1847. PD, WD, Ba.

Pascoe, Hodgson, 1829. Watchmaker and jeweller. 67 Chapel Street. Apprenticed to Jonathan Hodge of Helston. Settled at Penzance 1835. PD, WD, KD, KDD.

Pascoe, William, 1758. Silver pair-case watch. Ba.

LIST OF MAKERS

Quick, James, 1873. Watchmaker and jeweller. 8 Alverton Street. KDD.

Rickard, Hercules, 1812. Watch and clockmaker. RCG 18 Jul 1812, married to Miss Mary Sampson. (NB, no place mentioned, but Penzance district likely.)

Rossiter, Thomas, 1854. Watchmaker and jeweller. WB 6 Jan 1854, advert of watch and clock repair. Market Place, CDP. 2 Green Market, KDD. LC painted dial, walnut case, c. 1830.

Sampson, Henry, c. 1780. CC. Watchmaker. UBD, Ba. LC arch dial with face and moving eyes.

Sampson, John, 1752. Clockmaker and silversmith. SM 13 Apr 1752, wants 'any young man bred in the clock and watchmaking way wanting employment' to apply to him. Also apprentice. 1763, married Ann Leggo at Madron (parish Registers). 1762, wants apprentice; 1767, wants journeyman and apprentice. LC brass arch dial, Tempus Fugit in arch, 8-day, case softwood once japanned? c. 1775.

Scott, John, 1864. Watch and clock repairer. 31 Adelaide Street. CDP.

Selig, Benjamin, 1844. Watch and clockmaker. North Street. WB 8 Jun 1849, son born. PD, WD. Bracket clock, 8-day, mahogany case by Francis Rouse, St Ives.

Shortman, Samuel, 1823. Watch and clockmaker. East Street, pd, PD. 110 Market Jew Street, CDP, KD, wp. Bond 1852.

Sleeman, Henry, 1833. WB 2 Aug 1833, advertises for man who thoroughly understands clockmaking.

Solomon, Thomas, 1793. Watchmaker. Will, 1793.

Stephens, ——?, c. 1805. LC painted arch dial, moon-phases.

Tonkin, Thomas, 1768. Watchmaker. SM 5 Sep 1768, Gold watch, lost, with his name.

Trounson, William Henry, 1856. Watchmaker and jeweller. 31 Market Jew Street, KD, CDP. 3 Market Place, KDD.

Vibert, Charles, 1750-1809. Watchmaker. Will 1809. Market Jew Street house to wife Margaret and daughter Elizabeth. Another to daughter Mary Sparrow Vibert. LC brass arch dial, engraved centre, 8-day, c. 1790.

Vibert, John Pope, 1790-1865. Book and print seller, engraver, watches, clocks, etc. Market Jew Street. Nephew to above. Chapel Warden at St Mary's. Ba gives date 1780. pd, PD, WD, KD, wp. LC brass arch dial, flat, engraved, 8-day, c. 1820.

Vibert, Henry Pope, 1864. Jeweller, watchmaker, etc. 3 Market Place. CDP.

West, William, 1832. WB 30 Mar 1832, shop broken open, 2 watches stolen. (Possibly branch shop of WW, St Ives.)

PERRANUTHNOE

Thomas, Matthew Henry, 1844. Will 1844.

PROBUS

Hotton, William, 1827. Watchmaker, grocer, draper and postmaster. *Western Luminary,* 17 Apr 1827, theft of 17 watches. KD.

REDRUTH

Beringer, Schwerer & Co., 1844. Watch and clockmakers. Fore Street, PD, WD. 1 West End, KD, KDD. Still active as Beringer's.

Cohen, Emanuel, 1766-1849. Watch, clock and jewellery. Subscribed 5s in 1849 for starving Jews in Tiberias. Local tradition says he could be seen waiting for sunset on Friday and Saturday evenings to close and open shop. WB 22 Jun 1849, business to be disposed of after death of proprietor. Carried on for 30 years. PD. LC painted arch dial, 8-day, c. 1820. Silver pair-case watch, hallmark 1821.

Goldsworthy, Joseph, 1873. Watchmaker. East End. KDD.

Harris, George, 1818. Watchmaker. WB 6 Feb 1818, daring robbery on house; two silver watches and £21 stolen. LC brass arch plain dial, 8-day, c. 1820.

Harris, Thomas, 1823. Watchmaker. pd.

Hocking, John, 1844. Watch and clockmaker. Fore Street. PD, WD.

Hocking, Samuel, c. 1850. Watchmaker. Fore Street, south side.

Hocking, William, 1809. Churchwardens' a/cs 'For a wheel for the town clock £1-10s'.

Ivey, Nicholas, 1775. SM 14 Aug 1775, watch stolen from shop. LC brass arch dial, engraved spandrels, c. 1795.

Jacob, Moses, 1769. Watchmaker. James Dawson contracted to serve him as journeyman, 1769. One of the first to deal in mineralogical specimens. Will proved 1807, to son Levy Jacob all tools to be used by him. The will is signed in cursive Hebrew characters. RCG 6 Jun 1807, Sarah, widow, announces intention of carrying on business with help of son and journeyman. UBD. Ba. LC brass arch dial with rocking ship, 8-day, good movement, c. 1780. Another, with King Neptune riding the waves in the arch.

Joseph, Joseph, 1823. Watch and clockmaker and mineralogist. Fore Street. pd, PD, WD.

LIST OF MAKERS

Kistler, Thomas Andrews, 1873. Watchmaker and jeweller. Fore Street. KDD.

Nicholas, Joseph, 1821. Watchmaker. WB 1 Mar 1821, married to Hannah James of Penryn. (JN described as 'late of Redruth'.)

Nicholls, James Henry, 1856. Watch and clockmaker. Fore Street, KD. Cross Street, KDD.

Penrose, Richard, 1791. Watch and clockmaker, and Assay Master. Paid £3-3s for winding town clock for one year, 1806. RCG 17 Mar 1810, journeyman wanted. UBD, Ba.

Rice, Nicholas, 1823. Watchmaker. pd.

Thomas, John Marshall, 1844. Watch and clockmaker, etc. Fore Street, PD, WD. West End, KD, KDD. And at Penryn. LC silvered arch dial, spade hands, engraved motifs in spandrels, 8-day, c. 1850.

Trenerry, John, 1806. Married at Truro 1806 to Betsy Courtis Osler (parish Registers). RCG 12 Dec 1807, daring robbery from shop. Ba, under 'Trewery'. Wall clock, spring movement, fusee, brass circular dial, no striking train, c. 1835.

Trevena, William I, 1706-1786? Churchwarden in 1769. Among principal inhabitants of the town. Bankrupt 1775. LC brass dial, lacquer case, c. 1750.

Trevena, William II, 1799. Watchmaker. Fore Street. Rec £5-5s for winding town clock for one year. WB 11 Nov 1853, son dead. PD, KD, KDD. LC painted arch dial, 8-day, c. 1820.

ST AGNES

Bryant, John? LC, no details.

Hocking, Samuel, c. 1830. LC painted arch dial, 8-day.

Letcher, Walter, 1856. Watch and clockmaker. Churchtown. KD, KDD.

Menadue, William, 1844. Watch and clockmaker. PD. LC painted arch dial, 8-day, c. 1840. See Camborne.

Pearce, John, 1844. Watchmaker. PD.

Rowse, Thomas, 1852. Watch and clockmaker. Vicarage. WB 16 Jul 1852, married at Truro to Ann Ennor, both of St Agnes. KD, KDD.

Trevena, William, 1844. Watch and clockmaker. WB 12 Jun 1846, daughter born. PD. See Redruth.

ST ALLEN

Ivey, Henry, 1854. Watch and clockmaker. WB 28 Jul 1854, offers electro-galvanic machines and 'all kinds of elective apparatus made to order'.

ST AUSTELL

Andrew, Benjamin, 1755. Watch and clockmaker. Bond 1755. Jan 12 1762, 'To Andrews for repairs of a watch, 6s'. Trewithen MSS, Trewinnard Ledger. 1767, watch—his own make—stolen, and also in 1791. Small turret clock, single hand, at Trewithen, 'the great clock in the Eastern Office' cost £30, Bill paid 16 Dec 1759, Trewithen MSS. Will 1797. Shop and stock to son Benjamin. UBD, Ba. LC brass square dial, engraved centre, 15-day movement, mahogany case, c. 1790. LC brass arch dial, 8-day, c. 1780.

Baker, Jonas, c. 1770. Watchmaker, etc. SM 28 Aug 1775, lost watch to him. LC brass square dial, 30-hr, c. 1770.

Bennett, John, 1841. Watchmaker. Market Street. WB 16 Apr 1841, apprentice wanted, also in 1842. 10 May 1844, retiring from business; premises adjoin new market. PD.

Crapp, William, 1803. Clockmaker. Bond, Elizabeth, his widow.

Crowle, John, 1679. Paid 11s 'for keeping the clock and mending'. (Hammond, *A Cornish Parish,* p. 100.)

Edwards, Timothy, c. 1770. LC formerly 30-hr 10in. brass square dial, silvered chapter ring.

Francis, Andrew, 1806. Watchmaker. Married at St Austell. RCG 12 Apr 1806.

Glanvill, William, 1847. Watch and clockmaker, gunsmith and bellhanger. WD.

Hewetson, James, 1826. Watchmaker. Married at Creed to Eliza Teague, 17 Jan 1826.

Hext, William, 1743, 'undertook to repair and keep in order the Town Clock for seven years at two guineas per annum from Easter Monday last'. (Hammond, *A Cornish Parish,* p. 100.) From London. SM advert 17 May 1743, makes clocks, watches.

Higman, Jacob, 1763-1841. CC. Watch and clockmaker. Fore Street. WB 20 Dec 1811, robbery; 26 Feb 1823, apprentice wanted; 25 Nov 1825, journeyman wanted; 18 Sep 1835, burglary; 29 Oct 1841, Higman dead, aged 78. 19 Nov 1841, business for sale after 35 years. Ba, pd. LC brass arch dial engraved centre, 8-day, c. 1810.

Joseph, J., c. 1820. LC painted arch dial Arabic numerals, 8-day.

Long, John, 1847. Watch and clockmaker. WD.

Long, Robert, 1844. Watch and clockmaker. East Hill. PD.

Michael, Benjamin, 1768-1853. Watchmaker. Market Street. WB 19 Aug 1853, dead suddenly 'whilst on his road to Mevagissey, aged 85 years, one of the oldest inhabitants of the town'. LC painted square dial, Arabic numerals, 30-hr. 'Benjm. Michael, St Austell'. See Mevagissey.

Michael, Benjamin, 1800. Parish clerk and watchmaker. 25 Dec 1800, married (parish Registers). Probably same as above.

Michell, Benjamin & Co., 1847. Watch and clockmakers. WD.

Michell, William, 1844. Watch and clockmaker. Church Street. WB 4 Aug 1843, son born to Mr 'Mitchell'. PD.

Michael & Michell, c. 1840. LC painted arch dial, 8-day, good case.

Nicolls, Henry, 1791. Watch and clockmaker. SM 21 Jan 1793, wants apprentice. UBD.

Orchard, M., 1856. Watch and clockmaker. Church Street. KD.

Pearce, John, 1850. Watch and clockmaker. Fore Street. WB 4 Jan 1850, married Mary Crapp; 3 Dec 1852, apprentice wanted. KD.

Pentecost, William, 1799. Watchmaker. SM 25 Nov 1799. WB 25 Jan 1833, shop broken into. 1844, Hotel Road, PD.

Peter, T., 1856. Watch and clockmaker. Victoria Place. KD.

Petherick, William, 1781. Watchmaker. Watch with owner's name and date as numerals—'Henry Udy 1781'. Will 1784, substantial property.

Polkinghorne, Philip, 1764. EML. See under Truro, probably the same.

Rolling, William, 1839. Watchmaker. Will 1839. LC brass dial, 8-day.

Rosevear, John, 1788. Watch and clockmaker. Bond 1788. SM 11 Jun 1792, wants apprentice. Will 1798, all to wife. UBD, Ba. LC brass square dial, 30-hr.

Rowe (or Roe), James, 1823. Watchmaker. Fore Street. WB 1 Aug 1823, stolen watch offered to him. pd.

Spiegelhalter, Savory, 1873. Clock cleaner. 19 Duke Street. KDD.

Tallack, John, c. 1800. LC silvered brass square dial, 30-hr.

Truscott, Joseph, 1842. Jeweller, silversmith, watch and clockmaker. Fore Street. WB 18 Mar 1842, apprentice and journeyman wanted. Other adverts in 1844, 1853. PD, WD.

Wallis, Richard, 1748. Watchmaker. SM 7 Nov 1748, lost watch to be returned to him.

Wallis, William, 1741. Gunsmith and watchmaker. SM advert 15 Dec 1741, Will 1750. Tools etc to son Robert. LC brass arch dial with moving seasons in arch, c. 1745.

Whetter, Frederick, 1873. Watchmaker. 3 Western Hill. KDD.

Wills, Thomas, 1710. Clockmaker. Will 1739, house, shop to son Edmund, and £20 'to sett up his business'. LC brass square dial, single-hand, 30-hr, c. 1710. LC brass arch dial, 8-day, lacquer case, c. 1725.

ST BLAZEY

Trewin (sometimes Terwin), J., 1850. Watchmaker and general ironmonger. WB 18 Oct 1850, apprentice wanted. KD. LC painted arch dial, 8-day, c. 1830.

ST BREOKE

Geach, Thomas, 1799. Watchmaker. Leases tolls from the Wadebridge, with Edward Geach, sadler, 1799. Married Elizabeth Brewer at St Columb 26 Dec 1801 (parish Registers).

Treverton, John, 1728. Clockmaker. EML to Ann Harris of Jacobstow, 18 Sep 1728.

ST COLUMB

Couch, W. J., c. 1820. LC painted arch dial, 8-day.

Geach, Thomas, 1805. Watchmaker. RCG 5 Jan 1805, journeyman wanted. RCG 23 Jun 1809, house robbed. pd. See St Breoke. LC painted arch dial, 8-day, factory movement.

Jewell, W., c. 1820. LC painted arch dial, 8-day.

Key, William, 1809. RCG 22 Jul 1809, the Martyn business (qv) disposed of to William Key.

Leverton, W., 1837. Watchmaker. RCG 6 Oct 1837, dead.

Marshall, John, 1617. Holds a shop in consideration of repairing the bell collars and clock. (Green Book.)

Martyn, William, 1800. Watchmaker. Will 1809. RCG 17 Jun 1809, disposal of business. LC painted dial 8-day, c. 1800.

Oliver, ——?, 1791. Clockmaker. UBD, Ba.

Pye, John, 1740. Clockmaker. Will. Interesting inventory of goods.

Sampson, Richard, 1734-1814. Watchmaker. See Liskeard. LC brass square dial 11in. 8-day, c. 1780.

Teage, Thomas, 1585. Clockmaker. 'For hanginge of the firebell and keping of the Clocke, 6s 8d.' 1594, 'Paide to the clockemaker'. (Green Book.)

Truscott, James, 1847. Watch and clockmaker. Market Street. WD, KD, KDD. LC painted arch dial, 30-hr, c. 1830.

Walkey, John, c. 1800. Watchmaker. 1817, son baptised (parish Register). Chaise clock, small square verge movement with bridge cock, enamel dial, plain white, set in round mahogany case.

Webber, James, 1847. Watchmaker. Market Street. WB 15 Jan 1847, son born. KDD.

LIST OF MAKERS

Webber, John I, 1791. Watch and clockmaker. Married Jenifer Truscott, 1787 (parish Register). RCG 14 Jul 1804, 'dropt suddenly from his horse and died'. UBD, Ba. LC brass arch dial 'High Water at Padstow', 8-day, c. 1795.

Webber, John II, 1809. Watchmaker. RCG 23 Jun 1809, house broken open and robbed of several articles. 3 Feb 1810, journeyman wanted. WB 26 Jul 1811, in Bodmin Gaol as insolvent debtor. pd.

Webber, J., 1856. Watch and clockmaker. Bank Street. KD. This could be the above.

Webber, William, 1847. Watchmaker. WD.

ST COLUMB MINOR, NEWQUAY

Reynolds, J., 1851. Watchmaker. WB 20 Jun 1851, son born. KD.

ST DAY

Barratt, Peter, 1844. Watch and clockmaker. WB 8 Aug 1845, dead. PD.

Thomas, Richard, c. 1810. LC painted arch dial.

Veale, John, 1844. Watch and clockmaker. WB 20 Dec 1844, daughter born. PD, WD, KD, KDD. LC painted arch dial, 8-day.

Wilton, William, 1844. Mathematical instrument and watch and clockmaker. PD, WD. LC painted arch dial, 8-day, factory movement stamped 'Houghton', c. 1830.

ST DOMINIC

Westlake, ——?, c. 1840. LC painted arch dial, 30-hr, c. 1840.

ST ERTH

Dobb, H. (see Helston). LC painted arch dial, 8-day, c. 1830.

Wearn, Roger I, 1752. (OCW has Richard W, Camborne.) LC brass arch dial, engraved centre, dated 1752, 8-day.

Wearn, Roger II, 1748-1820. Watch and clockmaker. Son of above. Will 1820—tools etc to son Joseph; old bedroom clock to James; new 8-day to daughter Charity; new 30-hr to daughter Margaret. Epitaph in churchyard. LC brass arch dial, 'High Water at Hayle', 8-day, c. 1785.

ST GERMANS

Bassett, Benjamin, 1847. Watch and clockmaker. WD.

Tapson, ——?, 1829. Watch, hallmark 1829/30, engraved with name of owner, 'Rd Clemens St Keyne'.

ST GLUVIAS

Fonnereau, James, 1676. Watchmaker. Truro Lane. (Henderson MSS vol 8, no. 1328.)

ST IVES

Anthony, John Bray, 1844. Watch and clockmaker. Tregenna Place, PD, KD. And Registrar, Market Place, KDD.

Crouch, Thomas, before 1780. Ba.

Daniel, Matthew, 1873. Watchmaker. Tregenna Place. KDD.

Ley, John, 1805. Watchmaker. Market Place. Married Elizabeth Ellis 1805. RCG 9 Sep 1815, business to let, small stock; estab. 20 years.

Ninnis, Isaac, c. 1800. LC brass arch dial, 8-day.

Richards, ——?, 1787. LC brass arch dial, dated 1787, with minute ring arched between each 5-minute figure, 8-day.

Sandow, Thomas, 1844. Watch and clockmaker. Fore Street, PD, WD. 1856, as 'Sendrow', KD.

Stevens, John, 1823. Watchmaker. High Street. pd. LC painted arch dial, 8-day.

Stokes, John, before 1766. Ba.

West, William, 1751-1831. Watchmaker. Tregenna Place. Also Hayle and Helston. Married Joanna Harvey at St Erth in 1784. pd. LC brass arch dial, 'High Water at Hayle', c. 1810. LC with automata on painted arch dial, c. 1820.

Williams, Jasper, 1844. Watch and clockmaker. Fore Street West. PD, WD.

Wiseman, ——?, before 1752. Ba.

ST JUST-IN-PENWITH

Bottrall, Thomas, c. 1820. LC painted arch dial, 8-day.

Daniel, Nicholas Charles, 1847. Plumber, painter, glazier, tinplate worker and watchmaker. Fore Street. WD, KD, KDD.

Trounson, Thomas, 1873. Watchmaker and carpenter. Nancherrow Terrace. KDD.

ST KEYNE

Coryndon, William, 1765. Watchmaker. Late of Plymouth.

ST NEOT

Kernicke, Christopher, 1622. 'Pd for one dayes worke about the clock, 1s.' (Churchwardens a/cs.)

LIST OF MAKERS

Russell, Nicholas, 1609. 'Pd Nicholas Russell for coming to see the clocke—22 pence'; 1616, 'pd to Russel for mending the clock, 6s 6d'; 'pd the smith for Russell's working in his forge about the clock, 2d'. (Churchwardens a/cs.)

Sparrow, Thomas, 1722. Clockmaker.

SALTASH

Boney, Caleb, 1847. Watchmaker. Turnpike Gate. WD, KD. (Grandson of the Padstow Caleb I.)

Pearce, William Philip, 1873. Watch and clockmaker. 45 Fore Street. KDD.

Strathon, John Thomas, 1873. Watchmaker. 53 Fore Street. KDD.

STRATTON

Bevan, James, 1847. Watch and clockmaker. WD.

Bevan, Thomas, c. 1820. Watch and clockmaker. Business relinquished about 1845. PD. LC painted arch dial, moon-phase in arch, 8-day.

Ede, John, 1873. Watch and clockmaker. KDD.

Heckett (Hickett, Heket, Hyckyt, etc), Thomas, 1529. 'Pd to Thomas Hekyth to repair clok', High Cross Wardens' a/cs; 'Pd upto Thomas Heckett for the mending off the Cloke, xiijs iiijd', Blanchminster's Charity a/cs. 21 Nov 1538, married to Johanna Warmington at Poughill. Will 1572.

Mabyn, John, 1563. 'Payd unto John Mabyn for mending and kepyng off the cloke, vs', Blanchminster's Charity a/cs.

(Noted as keepers of the clock were William Archer, 1590; John Heckett, 1584?; Heddon, 1603; John Hocken, 1612; Davye Mabyn, 1601. The clock was new made 1562 by the 'clocke maker', again in 1594, and mended 1598 by unnamed craftsman, probably travelling clocksmiths.)

Mules, William, 1844. Watch and clockmaker. PD. See also Kilkhampton.

Petherick, John Cater, 1844. Watch and clockmaker. PD, WD.

Rattenbury, Frederick, 1847. Watch and clockmaker and emigration agent. WB 28 Mar 1851, son born. WD, KDD.

Uglow, George, 1791. Clockmaker. WB 19 Mar 1841, widow dead. UBD, Ba. LC painted square dial, 30-hr.

TORPOINT

Durant, John Pope, 1873. Watchmaker and jeweller. 52 Fore Street. KDD.

Ham, G., 1856. Watch and clockmaker. Fore Street. KD.

Soady, William Henry, 1873. Watchmaker. 20 Macey Street. KDD.

TREGONY

Eva, Richard, 1734-1806. Watch and clockmaker and silversmith. See Falmouth. Worked here for several years before moving to Falmouth abt 1780. Bond here 1776. Leading local Dissenter. Maker of town clock, Tregony, single-hand 30-hr. This clock, with other Borough property, was sold 1861 and was proposed to be sent to Australia. Townsfolk hid it until they were allowed to buy it. Survived as part of a pigsty till 1961, when it was sent to scrap just when it was about to be rescued. LC brass arch dial, engraved centre, 8-day. 'Richd Eva, Tregony', c. 1775.

Huddy, William Hotton, 1856. Watch and clockmaker and 'Gregor Arms', at Cornelly. Previously at Newlyn East. A maker. KD, KDD.

Turner, Thomas, c. 1790. LC brass arch dial engraved centre and spandrels, 8-day, c. 1790.

Williams, John, 1792. SM 22 Oct 1792, wants apprentice to clock and watchmaking.

TRURO

Anear, Frederick, 1834. Watchmaker. New Bridge Street. WB 25 Sep 1834, fire in house, in Kenwyn Street. PD, KD. Bellhanger, etc, KDD.

Anear, Charles Henry, 1873. Watchmaker. 104 Pydar Street. KDD.

Anthony, James, 1698. Borough Order Book, —negligent in keeping the Town Clock. Discharged, and John Thompson put in his room, 9 Dec 1698. James Anthony 'doe keepe the publick clock of this Bourrough, he makinge it a pendilow at his owne charge', £2-10s per annum, 4 Dec 1699.

Anthony, William, 1688-1768. House near the West Bridge. Will 1768. LC brass square dial, good hands, locking-plate striking, 8-day, c. 1730. SM 12 Aug 1765, watch lost, maker, 'Anthony, Truro'. 29 June 1767, another, 'God save the King' for the twelve hours.

Ball, Philip, 1769-1837. Watchmaker. 2 New Bridge Street. Constables' List 1803 shews him to be lame. RCG 15 May 1819, effects of Digory Wroath, a bankrupt, include 8-day clock by Ball. Churchwardens' a/cs

LIST OF MAKERS

10 May 1802, 'Kitto and Ball's bill for repairing the clock and altering the hammer, £1.' pd.

Ball, William, 1844. Watch and clockmaker. Quay Street. PD, WD.

Bennett, J., c. 1830. LC painted arch dial, 8-day, c. 1830.

Bennett, William, 1847. Watch and clockmaker. Duke Street. WD.

Beringer & Schwerer, 1849. Church Lane. WB 30 Nov 1849, journeyman watchmaker wanted.

Bunster, T., c. 1750. LC brass arch dial, moon-phase in arch. Repaired 1770.

Carkeek (or Karkeek), George, 1803. Watch and clockmaker. 9 High Cross. SM 12 Dec 1803, apprentice wanted—also be instructed in retail business of a silversmith, jeweller, cutler and ironmonger. RCG 21 Aug 1813, removing from 4 High Cross to 21 Boscawen Street. WB 7 Jun 1816, selling up, miscellaneous wares; only one clock. But, pd, 1823.

Collins, Richard, 1873. Watchmaker and jeweller. Victoria Place. KDD.

Davey, Ebenezer, 1823. Watchmaker. West Bridge. pd.

Dorrington, Theophilus Lutey, 1873. Watchmaker and jeweller. 11 Church Lane. KDD.

Floyd, Thomas, 1783-1842. Watch and clockmaker. Kenwyn Street. Constables' List, 1803. WB 14 Oct 1842, dead.

Furze, Gregory, c. 1795. Ba. LC brass arch dial, 8-day, c. 1795.

Geach, George, 1829. From London, also auctioneer and appraiser, WB 3 Jul 1829. 17 Jul 1829, offered stolen watch; 6 Jun 1834, dead. Succeeded by his son John H. Geach at 68 Pydar Street.

Gill, William, 1853. Watchmaker. WB 16 Dec 1853, son born.

Gubbin, John, 1823. Watchmaker. 32 St Nicholas Street. Will 1827. Stock-in-trade to wife Betsy. WB 30 Mar 1827, sale of business, 'working department equal to any in county'. pd.

Harris, George, 1822. Watchmaker. 4 Kenwyn Street. 21 Apr 1822, son baptised (parish Register). Also silversmith, ironmonger. pd, WD.

Harris, Henry, 1823. Watch and clockmaker. 2 Lemon Street, pd. 8 Lemon Street, PD. Also jeweller, silkmercer and haberdasher. WD. Formerly in Penzance.

Harris, Israel, 1844. Watch and clockmaker. 9 St Nicholas Street. PD.

Harris, John, 1847. Tea dealer and watchmaker. Boscawen Street. WD.

Hocking, James, 1856. Watch and clockmaker. 1 Castle Street. KD, KDD.

Husband, Walter, 1784. Borough a/cs 28 Jul 1784, 'Pd Walter Husband's bill for cleaning etc the tower clock (3 yrs) £7-10s'.

Ivey, Nicholas, 1791. Watchmaker. Bond. See Redruth.

Job, John Pentecost, 1805. Watchmaker. Kenwyn part of Truro. RCG 6 Jul 1805, watch lost—return to Mr Job. Bond to Elizabeth his mother, 1821.

Karkeek, George, 1823. See Carkeek.

Levy, Charles, 1844. Watch and clockmaker. 6 King Street. WB 11 Apr 1845, to open shop at 29 Boscawen Street; 23 Jan 1846, relinquishing shop in Boscawen Street and continuing at Kenwyn Street. Sale of ironmongery, toys, carpenters' and masons' tools, musical instruments, American and bracket clocks. PD.

Levy, Jacob, 1822. Watchmaker. King Street. 22 Feb 1822, four children baptised (parish Register). WD. LC painted arch dial, 8-day, c. 1820. pd, (St Nicholas Street).

Michell, Frederick Barron, 1873. Watchmaker, jeweller and ironmonger. 28 Boscawen Street. KDD.

Newton, Francis, 1787. Watchmaker. Kenwyn.

Pascoe, J., 1815. Watch and clock manufacturer, silversmith and jeweller. 18 Boscawen Street. RCG 2 Dec 1815, 'having been instructed by the first masters in finishing and making horiz. and other kinds of watches in the London stile, which is allowed to be superior to Country watch-makers in the general way—has commenced business'.

Pfaff, Joseph, c. 1800. Watch movement, enamel dial, verge escapement, c. 1800.

Phillips, Edwin, 1855. Watchmaker. Formerly of Wadebridge. WB 13 Jul 1855, son born.

Polkinghorne, Philip, 1764. EML. Died Truro 1801. LC fine case in mahogany, silvered brass arch dial, engraved centre, 8-day. See also St Austell.

Rossiter, Walter, 1823. Watchmaker. Bodmin Street, pd. 18 Feb 1827, two children baptised (parish Register). PD, WD, Kenwyn Street. 1856, St Mary's Street, KD.

Sampson, John, c. 1750. LC movement and dial, inner quarter markings, superior workmanship. Name and town on applied label—'Ino Sampson, Truroe', c. 1750.

Schwerer, J., 1856. Watch and clockmaker. Church Lane. KD.

Semmens, Herman, 1834. Watchmaker. New Bridge Street. WB 20 Jun 1834, daughter born. PD.

Sholl, Robert, 1765-1815. Watch and clockmaker. Market Place. His son, also watchmaker, became a deserter from the militia, 1811. Ba, under 'Shole'. Bracket clock, striking, plain silvered dial, engraved. Plain back plate, verge movement, 8-day. Good London-style case, c. 1790.

Skyrme, James, 1847. Watch and clockmaker. Kenwyn Street. WD, KD.

Thompson, John, 1698. Kept the public clock of Truro for one year.

Truscott, Lewis, 1833. Watchmaker and silversmith. St Nicholas Street. WB 13 Dec 1833, watch stolen. LC painted arch dial, 8-day, c. 1820.

Uglow, William, c. 1820. Watch and clockmaker. 8 King Street. RCG 25 Sep 1829, apprentice wanted, also journeyman in 1835 and 1846. PD, WD. LC brass arch dial, 8-day, c. 1820. 'Tempus Fugit' in arch.

Wallis, Richard, 1735. Died 1770 (parish Register). Ba. LC brass square dial, 8-day, c. 1770. Small turret movement, single hand, 8-day, well-made 4-post frame, minute dial on movement engraved 'Richard Wallis, Truro, June 1753'. Striking hours only.

Ward, Anthony, 1705. A/c book of Warwick Mohun of Luney in Creed, 'Mr Ward of Truroe' paid 2s 6d for putting clock right, and £5 7s 6d for a new watch. OCW. LC month movement, brass square dial, herring-bone border, etc, c. 1705.

(An Anthony Ward reported in Philadelphia in 1717, and six clocks still extant with this name in handsome American walnut cases. Supposed to have died New York 1731-35.)

Watson, ——?, 1681. Truro Borough Minute Book, paid 12s 6d for keeping the clock, and 5s towards present repairs. He undertook to 'keepe the clock' at 50s per annum.

Weisbarth, Charles William, 1873. Watchmaker. 16 Boscawen Street. KDD.

Wills, Richard, 1752. Watchmaker. In business in main street from about 1750-1805. EML. Ba. Cescinsky and Webster, *English Domestic Clocks*, illustration. Tower clock at St Mary's, Truro, 1770. LC brass arch dial, with moon-phase and 'High Water at Truro Key', 30-hr with carillon on eight bells every four hours, changing automatically each time. Sundial, Wendron Church, 1770, etc. A fine maker with superior standards.

Wills, William, 1756-1819. Watchmaker. Son of above. 1784, leased St George's Park, in Lostwithiel. SM 21 Jun 1802, apprentice wanted; will be taught gunsmith's work as well. Bond 1820 to Ann his sister. Completed astronomical clock with musical (organ) train (perhaps begun by father) open to public subscription. LC brass arch dial, engraved centre, 8-day, c. 1800. LC painted arch dial, 8-day, with swans feeding in arch with pendulum swing, poor.

WADEBRIDGE

Billing, Chalwell, 1802. Watchmaker. SM 15 Nov 1802, journeyman wanted; 12 Mar 1804, declining business—stock of watches, plate, jewellery, ironmongery, etc. RCG 8 Sep 1804, sale of bankrupt's effects.

Broad, John Butler, 1856. Watch and clockmaker. KD, KDD.

Reynolds, James, 1844. Watch and clockmaker. Noted under Bodmin in PD.

Reynolds, John, c. 1780. LC brass square dial, 8-day. See Egloshayle.

Reynolds, J., 1856. Watch and clockmaker. KD.

Reynolds, Thomas, 1786. Watchmaker. Wadebridge a/cs—Trustees lease to him a dwelling-house built by Humphry Gink, dec'd, on the east side of Wadebridge 47 ft by 14 ft. LC brass square dial, 8-day.

WITHIEL

Pedler, Joseph, born about 1739. Considerable business as smith at Withiel-goose. Made curious clocks. LC brass arch dial, oak case, c. 1770.

SHORT BIBLIOGRAPHY

The literature of horology is extensive, but a few books are suggested here for the reader who desires an introduction to the general subject of clocks and clockmaking in this country.

Britten, F. J.	*Old Clocks and Watches and their Makers*. Seventh ed by G. H. Baillie, C. Clutton and C. A. Ilbert. 1956.
Cescinsky, H. and Webster, M.	*English Domestic Clocks*. 1913.
Lloyd, H. Alan	*Chats on Old Clocks*. 1951.
„	*Old Clocks*. Second ed. 1958.
Rees, Abraham	*Clocks Watches and Chronometers*. (A Selection from Rees's Cyclopaedia, 1819-20). David & Charles. 1970.
Symonds, R. W.	*A Book of English Clocks*. King Penguin Series. 1947.
Ullyett, K.	*British Clocks and Clockmakers*. 1947.
„	*In Quest of Clocks*. 1951.